About Island Press

Island Press is the only nonprofit organization in the United States whose principal purpose is the publication of books on environmental issues and natural resource management. We provide solutions-oriented information to professionals, public officials, business and community leaders, and concerned citizens who are shaping responses to environmental problems.

In 2003, Island Press celebrates its nineteenth anniversary as the leading provider of timely and practical books that take a multidisciplinary approach to critical environmental concerns. Our growing list of titles reflects our commitment to bringing the best of an expanding body of literature to the environmental community throughout North America and the world.

Support for Island Press is provided by The Nathan Cummings Foundation, Geraldine R. Dodge Foundation, Doris Duke Charitable Foundation, Educational Foundation of America, The Charles Engelhard Foundation, The Ford Foundation, The George Gund Foundation, The Vira I. Heinz Endowment, The William and Flora Hewlett Foundation, Henry Luce Foundation, The John D. and Catherine T. MacArthur Foundation, The Andrew W. Mellon Foundation, The Moriah Fund, The Curtis and Edith Munson Foundation, National Fish and Wildlife Foundation, The New-Land Foundation, Oak Foundation, The Overbrook Foundation, The David and Lucile Packard Foundation, The Pew Charitable Trusts, The Rockefeller Foundation, The Winslow Foundation, and other generous donors.

The opinions expressed in this book are those of the author(s) and do not necessarily reflect the views of these foundations.

Ecosystems and Human Well-being

A Report of the Conceptual Framework Working Group
of the Millennium Ecosystem Assessment

Millennium Ecosystem Assessment Board

The MA Board represents the users of the findings of the MA process.

Co-chairs

Robert T. Watson, *World Bank*
A.H. Zakri, *United Nations University*

Institutional Representatives

Delmar Blasco, *Ramsar Convention on Wetlands*
Peter Bridgewater, *United Nations Educational, Scientific and Cultural Organization*
Philbert Brown, *Convention to Combat Desertification*
Hama Arba Diallo, *Convention to Combat Desertification*
Max Finlayson, *Ramsar Convention on Wetlands*
Colin Galbraith, *Convention on Migratory Species*
Richard Helmer, *World Health Organization*
Yolanda Kakabadse, *World Conservation Union*
Arnulf Müller-Helmbrecht, *Convention on Migratory Species*
Alfred Oteng-Yeboah, *Convention on Biological Diversity*
Seema Paul, *United Nations Foundation*
Mario Ramos, *Global Environment Facility*
Thomas Rosswall, *International Council for Science*
Dennis Tirpak, *Framework Convention on Climate Change*
Klaus Töpfer, *United Nations Environment Programme*
Jeff Tschirley, *Food and Agriculture Organization of the United Nations*
Alvaro Umaña, *United Nations Development Programme*
Meryl Williams, *Consultative Group on International Agricultural Research*
Hamdallah Zedan, *Convention on Biological Diversity*

At-large Members

Fernando Almeida	José María Figueres	Paul Maro	Ismail Serageldin
Phoebe Barnard	Fred Fortier	Hal Mooney	David Suzuki
Gordana Beltram	Mohammed H.A. Hassan	Marina Motovilova	M.S. Swaminathan
Antony Burgmans	Yoriko Kawaguchi	M.K. Prasad	José Tundisi
Esther Camac	Corinne Lepage	Walter V. Reid	Axel Wenblad
Angela Cropper	Jonathan Lash	Henry Schacht	Xu Guanhua
Partha Dasgupta	Wangari Maathai	Peter Johan Schei	Muhammad Yunus

Millennium Ecosystem Assessment Secretariat

The United Nations Environment Programme (UNEP) coordinates the Millennium Ecosystem Assessment Secretariat, which is based at the following partner institutions:

Food and Agriculture Organization of the United Nations (FAO), Italy
Institute of Economic Growth, India
Meridian Institute, USA
National Institute of Public Health and the Environment (RIVM), Netherlands
Scientific Committee on Problems of the Environment (SCOPE), France
UNEP-World Conservation Monitoring Centre, United Kingdom
University of Pretoria, South Africa
University of Wisconsin, USA
World Resources Institute (WRI), USA
WorldFish Center, Malaysia

Ecosystems and Human Well-being: A Framework for Assessment

Authors

Joseph Alcamo
Neville J. Ash
Colin D. Butler
J. Baird Callicott
Doris Capistrano
Stephen R. Carpenter
Juan Carlos Castilla
Robert Chambers
Kanchan Chopra
Angela Cropper
Gretchen C. Daily
Partha Dasgupta
Rudolf de Groot
Thomas Dietz
Anantha Kumar Duraiappah
Madhav Gadgil
Kirk Hamilton

Rashid Hassan
Eric F. Lambin
Louis Lebel
Rik Leemans
Liu Jiyuan
Jean-Paul Malingreau
Robert M. May
Alex F. McCalla
Tony (A.J.) McMichael
Bedrich Moldan
Harold Mooney
Shahid Naeem
Gerald C. Nelson
Niu Wen-Yuan
Ian Noble
Ouyang Zhiyun
Stefano Pagiola

Daniel Pauly
Steve Percy
Prabhu Pingali
Robert Prescott-Allen
Walter V. Reid
Taylor H. Ricketts
Cristian Samper
Robert (Bob) Scholes
Henk Simons
Ferenc L. Toth
Jane K. Turpie
Robert Tony Watson
Thomas J. Wilbanks
Meryl Williams
Stanley Wood
Zhao Shidong
Monika B. Zurek

Contributing Authors

Elena M. Bennett
Reinette (Oonsie) Biggs
Poh-Sze Choo
Jonathan Foley
Pushpam Kumar
Marcus J. Lee
Richard H. Moss
Gerhard Petschel-Held
Sarah Porter
Stephen H. Schneider

Assessment Panel Chairs

Angela Cropper
Harold A. Mooney

MA Director

Walter V. Reid

Editorial Board Chairs

José Sarukhán
Anne Whyte

Chapter Review Editors

Gilberto Gallopin
Roger Kasperson
Mohan Munasinghe
Léon Olivé
Christine Padoch
Jeffrey Romm
Hebe Vessuri

ISLAND PRESS

Washington • Covelo • London

ISLAND PRESS is a trademark of The Center for Resource Economics.

Library of Congress Cataloging-in-Publication Data

Ecosystems and human well-being : a framework for assessment /
Millennium Ecosystem Assessment ; authors, Joseph Alcamo [et al.] ;
contributing authors, Elena M. Bennett [et al.].
     p. cm.

"The first product of the Millennium Ecosystem Assessment (MA), a
four-year international work program designed to meet the needs of
decision-makers for scientific information on the links between
ecosystem change and human well-being"—Pref.
Includes bibliographical references and index.
ISBN 1-55963-402-2 (cloth : alk. paper) —
ISBN 1-55963-403-0 (pbk. : alk. paper)

1. Human ecology. 2. Ecosystem management. I. Alcamo, Joseph. II.
Bennett, Elena M. III. Millennium Ecosystem Assessment (Program)
GF50.E26 2003
333.95—dc21

                     2003011612

British Cataloguing-in-Publication Data available

Printed on recycled, acid-free paper ⊛
Manufactured in the United States of America

09 08 07 06 05 04 03   10 9 8 7 6 5 4 3 2 1

The Board of the Millennium Ecosystem Assessment dedicates this report to the memory of Angela Cropper's husband, mother, and sister: John Cropper, Maggie Lee and Lynette Lithgow-Pearson. Through their lives and work they embodied the spirit and intent of the Millennium Ecosystem Assessment by their love of the natural world and their concern to improve the lives of people.

Table of Contents

Preface

Ecosystems and Human Well-being: A Framework for Assessment is the first product of the Millennium Ecosystem Assessment (MA), a four-year international work program designed to meet the needs of decision-makers for scientific information on the links between ecosystem change and human well-being. It was launched by United Nations Secretary-General Kofi Annan in June 2001, and the principal assessment reports will be released in 2005. The MA focuses on how changes in ecosystem services have affected human well-being, how ecosystem changes may affect people in future decades, and what types of responses can be adopted at local, national, or global scales to improve ecosystem management and thereby contribute to human well-being and poverty alleviation.

Parties to the Convention on Biological Diversity, the Convention to Combat Desertification, the Ramsar Convention on Wetlands, and the Convention on Migratory Species have asked the MA to provide scientific information to assist in the implementation of these treaties. The MA will also address the needs of other stakeholders, including the private sector, civil society, and indigenous peoples organizations. The MA is closely coordinated with other international assessments that focus in greater depth on particular sectors or drivers of change, such as the Intergovernmental Panel on Climate Change and the Global International Waters Assessment. Scientific evaluations such as these help underpin various regular annual and biennial international reporting mechanisms, such as the *Global Environmental Outlook*, the *World Resources Report*, the *Human Development Report*, and the *World Development Report*.

Leading scientists from more than 100 nations are conducting the MA under the direction of a Board that includes representatives of five international conventions, five United Nations agencies, international scientific organizations, and leaders from the private sector, nongovernmental organizations, and indigenous groups. If the MA proves to be useful to its stakeholders, it is anticipated that an integrated ecosystem assessment process modeled on this process will be repeated at a global scale every 5–10 years and that ecosystem assessments will be regularly conducted at national or sub-national scales.

An ecosystem assessment can aid any country, region, or company by:

- deepening understanding of the relationship and linkages between ecosystems and human well-being;
- demonstrating the potential of ecosystems to contribute to poverty reduction and enhanced well-being;
- evaluating the compatibility of policies established by institutions at different scales;

- integrating economic, environmental, social, and cultural aspirations;
- integrating information from both natural and social science;
- identifying and evaluating policy and management options for sustaining ecosystem services and harmonizing them with human needs; and
- facilitating integrated ecosystem management.

The MA will help both in choosing among existing options and in identifying new approaches to carrying out the Plan of Implementation adopted at the World Summit on Sustainable Development (WSSD) and achieving the United Nations Millennium Development Goals. The WSSD Plan reiterates those goals and states that in order to "reverse the current trend in natural resource degradation as soon as possible, it is necessary to implement strategies which should include targets adopted at the national and, where appropriate, regional levels to protect ecosystems and to achieve integrated management of land, water and living resources, while strengthening regional, national and local capacities."

The MA will contribute directly to this goal and can respond to the WSSD call to:

> improve policy and decision-making at all levels through, *inter alia*, improved collaboration between natural and social scientists, and between scientists and policy makers, including through urgent actions at all levels to: (a) Increase the use of scientific knowledge and technology, and increase the beneficial use of local and indigenous knowledge in a manner respectful of the holders of that knowledge and consistent with national law; (b) Make greater use of integrated scientific assessments, risk assessments and interdisciplinary and intersectoral approaches;... .

The MA also seeks to help build individual and institutional capacity to undertake integrated ecosystem assessments and to act on their findings. In the final analysis, societies need to be enabled to manage their biological resources and their ecosystems better with the resources at hand. The human capacity to do so is vital. Wherever the MA activities unfold, they will leave a corps of more aware and motivated collaborators to continue the effort to achieve more enlightened and effective management.

This first report of the Millennium Ecosystem Assessment describes the conceptual framework that is being used in the MA. It is not a formal assessment of the literature, but rather a scientifically informed presentation of the choices made by the assessment team in structuring the analysis and framing the issues. The conceptual framework elaborated in this report describes the approach and assumptions that will underlie the analysis conducted in the Millennium Ecosystem Assessment. The framework was developed through interactions among the experts involved in the MA as well as stakeholders who will use its findings. It represents one means of examining the linkages between ecosystems and human well-being that is both scientifically credible and relevant to decision-makers. This framework for analysis and decision-making should be of use to a wide array of individuals and institutions in government, the private sector, and civil society that seek to incorporate considerations of ecosystem services in their assessments, plans, and actions.

Five overarching questions, along with the detailed lists of user needs provided by convention secretariats and the private sector, guide the issues being assessed:

- What are the current conditions and trends of ecosystems and their associated human well-being?
- What are the plausible future changes in ecosystems and in the supply of and demand for ecosystem services and the consequent changes in health, livelihood, security, and other constituents of well-being?
- What can we do to enhance well-being and conserve ecosystems? What are the strengths and weaknesses of response options, actions, and processes that can be considered to realize or avoid specific futures?
- What are the most robust findings and key uncertainties that affect the provision of ecosystem services (including the consequent changes in health, livelihood, and security) and other management decisions and policy formulations?
- What tools and methodologies developed and used in the MA can strengthen capacity to assess ecosystems, the services they provide, their impacts on human well-being, and the implications of response options?

The MA was launched in June 2001, and the final global assessment reports will be released in 2005. In addition, a series of short synthesis reports will be prepared, targeted at the needs of specific audiences, including the international conventions and the private sector. Up to 15 sub-global assessments may be carried out at local, national, and regional scales using this same conceptual framework and designed to contribute to decision-making at those scales. These sub-global assessments have already begun to release initial findings and will continue through 2006. During the course of the assessments, an ongoing dialogue is under way with the users at global and sub-global scales in order to ensure that the assessments are responsive to the needs of the users and that the users are informed regarding the potential utility of the findings.

This report has undergone two rounds of peer-review, first by experts involved in other parts of the MA process and then by both experts and governments (through the national focal points of the Convention on Biological Diversity, Convention to Combat Desertification, and the Ramsar Convention on Wetlands and through participating National Academies of Science).

Acknowledgments

The conceptual framework for the Millennium Ecosystem Assessment (MA) has been shaped by a large number of people since 1998, including the MA Exploratory Steering Committee, the MA Board, and the participants in two design meetings in 2001 (Netherlands and South Africa). We would particularly like to acknowledge the support and guidance from the scientific and technical bodies of the Convention on Biological Diversity (CBD), the Ramsar Convention on Wetlands, and the Convention to Combat Desertification (CCD), which have helped to define the focus of the MA.

We would like to acknowledge the contributions of all of the authors of this book, and the support provided by their institutions that enabled their participation. We would like to thank the MA Secretariat and the host organizations of the MA Technical Support Units—the WorldFish Center (Malaysia); the UNEP-World Conservation Monitoring Centre (United Kingdom); the Institute of Economic Growth (India); National Institute of Public Health and the Environment (RIVM) (Netherlands); the World Resources Institute, the Meridian Institute, and the Center for Limnology, University of Wisconsin (United States); the Scientific Committee on Problems of the Environment (France); and the International Maize and Wheat Improvement Center (CIMMYT) (Mexico)—for the support they provided in the preparation of this report. We thank several individuals who played particularly critical roles: Sara Suriani, Christine Jalleh, and Laurie Neville for their administrative and logistical support to the preparation of the report, Linda Starke for editing the report, Lori Han and Carol Rosen for managing the production process, and Maggie Powell for the preparation of the figures and final text. And, we thank past members of the MA Board whose contributions were instrumental in shaping the MA focus and process, including Gisbert Glaser, He Changchui, Ann Kern, Roberto Lenton, Hubert Markl, Susan Pineda Mercado, Jan Plesnik, Peter Raven, Cristian Samper, and Ola Smith. We also thank the individuals, institutions, and governments that submitted review comments on drafts of this report (listed in Appendix 2).

Financial support for the MA and the MA Sub-global Assessments is being provided by the Global Environment Facility (GEF), the United Nations Foundation, The David and Lucile Packard Foundation, The World Bank, the United Nations Environment Programme (UNEP), the Government of Norway, the Kingdom of Saudi Arabia, the Swedish International Biodiversity Programme, The Rockefeller Foundation, the United States National Aeronautic and Space Administration (NASA), the International Council for Science (ICSU), the Asia Pacific Network for Global Change Research, The Christensen Fund, the United Kingdom Department for Environment, Food and Rural Affairs (DEFRA), the Consultative Group for International Agricultural Research (CGIAR), and The

Ford Foundation. Generous in-kind support for the MA has been provided by the United Nations Development Programme (UNDP), the United Nations Educational Scientific and Cultural Organization (UNESCO), the Food and Agriculture Organization of the United Nations (FAO), the World Health Organization (WHO), the WorldFish Center, the Government of China, the Government of Germany, the Japan Ministry of Environment, the Asia Pacific Environmental Innovation Strategy Project (APEIS), the World Agroforestry Centre (ICRAF), Stockholm University, the Government of India, the Tropical Resources Ecology Program (TREP) of the University of Zimbabwe, the Department of Environment and Natural Resources of the Philippines, the Coast Information Team of British Columbia, Canada, and a large number of institutions that have supported staff time and travel. (A complete list of donors is available at http://www.millenniumassessment.org.)

The work to establish and design the MA was supported by grants from The Avina Group, The David and Lucile Packard Foundation, GEF, the Government of Norway, the Swedish International Development Cooperation Authority (SIDA), The Summit Foundation, UNDP, UNEP, the United Nations Foundation, the United States Agency for International Development (USAID), the Wallace Global Fund, and The World Bank.

Summary

Human well-being and progress toward sustainable development are vitally dependent upon improving the management of Earth's ecosystems to ensure their conservation and sustainable use. But while demands for ecosystem services such as food and clean water are growing, human actions are at the same time diminishing the capability of many ecosystems to meet these demands. Sound policy and management interventions can often reverse ecosystem degradation and enhance the contributions of ecosystems to human well-being, but knowing when and how to intervene requires substantial understanding of both the ecological and the social systems involved. Better information cannot guarantee improved decisions, but it is a prerequisite for sound decision-making.

The Millennium Ecosystem Assessment (MA) will help provide the knowledge base for improved decisions and will build capacity for analyzing and supplying this information. This document presents the conceptual and methodological approach that the MA will use to assess options that can enhance the contribution of ecosystems to human well-being. This same approach should provide a suitable basis for governments, the private sector, and civil society to factor considerations of ecosystems and ecosystem services into their own planning and actions.

Humanity has always depended on the services provided by the biosphere and its ecosystems. Further, the biosphere is itself the product of life on Earth. The composition of the atmosphere and soil, the cycling of elements through air and waterways, and many other ecological assets are all the result of living processes—and all are maintained and replenished by living ecosystems. The human species, while buffered against environmental immediacies by culture and technology, is ultimately fully dependent on the flow of ecosystem services.

In his April 2000 Millennium Report to the United Nations General Assembly, in recognition of the growing burden that degraded ecosystems are placing on human well-being and economic development and the opportunity that better managed ecosystems provide for meeting the goals of poverty eradication and sustainable development, United Nations Secretary-General Kofi Annan stated that:

> It is impossible to devise effective environmental policy unless it is based on sound scientific information. While major advances in data collection have been made in many areas, large gaps in our knowledge remain. In particular, there has never been a comprehensive global assessment of the world's major ecosystems. The planned Millennium Ecosystem Assessment, a major international collaborative effort to map the health of our planet, is a response to this need.

The Millennium Ecosystem Assessment was established with the involvement of governments, the private sector, nongovernmental organizations, and scientists to provide an integrated assessment of the consequences of ecosystem change for human well-being and to analyze options available to enhance the conservation of ecosystems and their contributions to meeting human needs. The Convention on Biological Diversity, the Convention to Combat Desertification, the Convention on Migratory Species, and the Ramsar Convention on Wetlands plan to use the findings of the MA, which will also help meet the needs of others in government, the private sector, and civil society. The MA should help to achieve the United Nations Millennium Development Goals and to carry out the Plan of Implementation of the 2002 World Summit on Sustainable Development. It will mobilize hundreds of scientists from countries around the world to provide information and clarify science concerning issues of greatest relevance to decision-makers. The MA will identify areas of broad scientific agreement and also point to areas of continuing scientific debate.

The assessment framework developed for the MA offers decision-makers a mechanism to:

- **Identify options that can better achieve core human development and sustainability goals.** All countries and communities are grappling with the challenge of meeting growing demands for food, clean water, health, and employment. And decision-makers in the private and public sectors must also balance economic growth and social development with the need for environmental conservation. All of these concerns are linked directly or indirectly to the world's ecosystems. The MA process, at all scales, will bring the best science to bear on the needs of decision-makers concerning these links between ecosystems, human development, and sustainability.

- **Better understand the trade-offs involved—across sectors and stakeholders—in decisions concerning the environment.** Ecosystem-related problems have historically been approached issue by issue, but rarely by pursuing multisectoral objectives. This approach has not withstood the test of time. Progress toward one objective such as increasing food production has often been at the cost of progress toward other objectives such as conserving biological diversity or improving water quality. The MA framework complements sectoral assessments with information on the full impact of potential policy choices across sectors and stakeholders.

- **Align response options with the level of governance where they can be most effective.** Effective management of ecosystems will require actions at all scales, from the local to the global. Human actions now directly or inadvertently affect virtually all of the world's ecosystems; actions required for the management of ecosystems refer to the steps that humans can take to modify their direct or indirect influences on ecosystems. The management and policy options available and the concerns of stakeholders differ greatly across these scales. The priority areas for biodiversity conservation in a country as defined based on "global" value, for example, would be very different from those as defined based on the value to local communities. The multiscale assessment framework developed for the MA provides a new approach for analyzing policy options at all scales—from local communities to international conventions.

What Is the Problem?

Ecosystem services are the benefits people obtain from ecosystems, which the MA describes as provisioning, regulating, supporting, and cultural services. (See Box 1.) Ecosystem services include products such as food, fuel, and fiber; regulating services such as climate regulation and disease control; and nonmaterial benefits such as spiritual or aesthetic benefits.

BOX 1.   Key Definitions

Ecosystem. An ecosystem is a dynamic complex of plant, animal, and microorganism communities and the nonliving environment interacting as a functional unit. Humans are an integral part of ecosystems. Ecosystems vary enormously in size; a temporary pond in a tree hollow and an ocean basin can both be ecosystems.

Ecosystem services. Ecosystem services are the benefits people obtain from ecosystems. These include provisioning services such as food and water; regulating services such as regulation of floods, drought, land degradation, and disease; supporting services such as soil formation and nutrient cycling; and cultural services such as recreational, spiritual, religious and other nonmaterial benefits.

Well-being. Human well-being has multiple constituents, including basic material for a good life, freedom and choice, health, good social relations, and security. Well-being is at the opposite end of a continuum from poverty, which has been defined as a "pronounced deprivation in well-being." The constituents of well-being, as experienced and perceived by people, are situation-dependent, reflecting local geography, culture, and ecological circumstances.

Changes in these services affect human well-being in many ways. (See Figure 1.)

The demand for ecosystem services is now so great that trade-offs among services have become the rule. A country can increase food supply by converting a forest to agriculture, for example, but in so doing it decreases the supply of services that may be of equal or greater importance, such as clean water, timber, ecotourism destinations, or flood regulation and drought control. There are many indications that human demands on ecosystems will grow still greater in the coming decades. Current estimates of 3 billion more people and a quadrupling of the world economy by 2050 imply a formidable increase in demand for and consumption of biological and physical resources, as well as escalating impacts on ecosystems and the services they provide.

The problem posed by the growing demand for ecosystem services is compounded by increasingly serious degradation in the capability of ecosystems to provide these services. World fisheries are now declining due to overfishing, for instance, and some 40 percent of agricultural land has been degraded in the past half-century by erosion, salinization, compaction, nutrient depletion, pollution, and urbanization. Other human-induced impacts on ecosystems include alteration of the nitrogen, phosphorous, sulfur, and carbon cycles, causing acid rain, algal blooms, and fish kills in rivers and coastal waters, along with contributions to climate change. In many parts of the world, this degradation of ecosystem services is exacerbated by the associated loss of the knowledge and understanding held by local communities—knowledge that sometimes could help to ensure the sustainable use of the ecosystem.

This combination of ever-growing demands being placed on increasingly degraded ecosystems seriously diminishes the prospects for sustainable development. Human well-being is affected not just by gaps between ecosystem service supply and demand but also by the increased vulnerability of individuals, communities, and nations. Productive ecosystems, with their array of services, provide people and communities with resources and options they can use as insurance in the face of natural catastrophes or social upheaval. While well-managed ecosystems reduce risks and vulnerability, poorly managed systems can exacerbate them by increasing risks of flood, drought, crop failure, or disease.

Ecosystem degradation tends to harm rural populations more directly than urban populations and has its most direct and severe impact on poor people. The wealthy control access to a greater share of ecosystem services, consume those services at a higher per capita rate, and are buffered

FIGURE 1.   Ecosystem Services and Their Links to Human Well-being

Ecosystem services are the benefits people obtain from ecosystems. These include provisioning, regulating, and cultural services, which directly affect people, and supporting services needed to maintain the other services. Changes in these services affect human well-being through impacts on security, the basic material for a good life, health, and social and cultural relations. These constituents of well-being are, in turn, influenced by and have an influence on the freedoms and choices available to people.

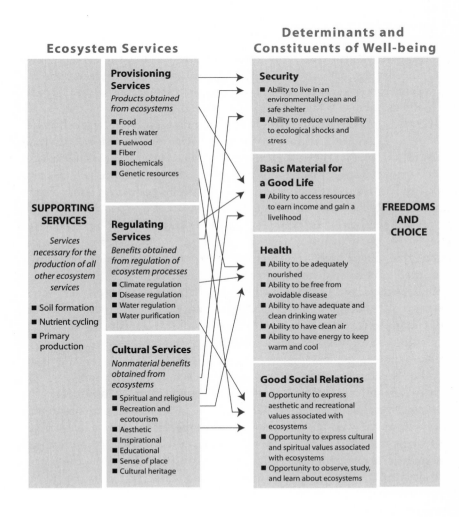

from changes in their availability (often at a substantial cost) through their ability to purchase scarce ecosystem services or substitutes. For example, even though a number of marine fisheries have been depleted in the past century, the supply of fish to wealthy consumers has not been disrupted since fishing fleets have been able to shift to previously underexploited stocks. In contrast, poor people often lack access to alternate services and are highly vulnerable to ecosystem changes that result in famine, drought, or floods. They frequently live in locations particularly sensitive to environmental threats, and they lack financial and institutional buffers against these dangers. Degradation of coastal fishery resources, for instance, results in a decline in protein consumed by the local community since fishers may not have access to alternate sources of fish and community members may not have enough income to purchase fish. Degradation affects their very survival.

Changes in ecosystems affect not just humans but countless other species as well. The management objectives that people set for ecosystems and the actions that they take are influenced not just by the consequences of ecosystem changes for humans but also by the importance people place on considerations of the intrinsic value of species and ecosystems. Intrinsic value is the value of something in and for itself, irrespective of its utility for someone else. For example, villages in India protect "spirit sanctuaries" in relatively natural states, even though a strict cost-benefit calculation might favor their conversion to agriculture. Similarly, many countries have passed laws protecting endangered species based on the view that these species have a right to exist, even if their protection results in net economic costs. Sound ecosystem management thus involves steps to address the utilitarian links of people to ecosystems as well as processes that allow considerations of the intrinsic value of ecosystems to be factored into decision-making.

The degradation of ecosystem services has many causes, including excessive demand for ecosystem services stemming from economic growth, demographic changes, and individual choices. Market mechanisms do not always ensure the conservation of ecosystem services either because markets do not exist for services such as cultural or regulatory services or, where they do exist, because policies and institutions do not enable people living within the ecosystem to benefit from services it may provide to others who are far away. For example, institutions are now only beginning to be developed to enable those benefiting from carbon sequestration to provide local managers with an economic incentive to leave a forest uncut, while strong economic incentives often exist for managers to harvest the forest. Also, even if a market exists for an ecosystem service, the results obtained through the market may be socially or ecologically undesirable. Properly managed, the creation of ecotourism opportunities in a country can create strong economic incentives for the maintenance of the cultural services provided by ecosystems, but poorly managed ecotourism activities can degrade the very resource on which they depend. Finally, markets are often unable to address important intra- and intergenerational equity issues associated with managing ecosystems for this and future generations, given that some changes in ecosystem services are irreversible.

The world has witnessed in recent decades not just dramatic changes to ecosystems but equally profound changes to social systems that shape both the pressures on ecosystems and the opportunities to respond. The relative influence of individual nation-states has diminished with the growth of power and influence of a far more complex array of institutions, including regional governments, multinational companies, the United Nations, and civil society organizations. Stakeholders have become more involved in decision-making. Given the multiple actors whose decisions now strongly influence ecosystems, the challenge of providing information to decision-makers has grown. At the same time, the new institutional landscape may provide an unprecedented opportunity for information concerning ecosystems to make a major difference. Improvements in ecosystem management to enhance human well-being will require new institutional and policy arrangements and changes in rights and access to resources that may be more possible today under these conditions of rapid social change than they have ever been before.

Like the benefits of increased education or improved governance, the protection, restoration, and enhancement of ecosystem services tends to have multiple and synergistic benefits. Already, many governments are beginning to recognize the need for more effective management of these basic life-support systems. Examples of significant progress toward sustainable management of biological resources can also be found in civil society, in indigenous and local communities, and in the private sector.

Conceptual Framework

The conceptual framework for the MA places human well-being as the central focus for assessment, while recognizing that biodiversity and ecosystems also have intrinsic value and that people take decisions concerning ecosystems based on considerations of well-being as well as intrinsic value. (See Box 2.) The MA conceptual framework assumes that a dynamic interaction exists between people and ecosystems, with the changing human condition serving to both directly and indirectly drive change in ecosystems and with changes in ecosystems causing changes in human well-being. At the same time, many other factors independent of the environment change the human condition, and many natural forces are influencing ecosystems.

The MA focuses particular attention on the linkages between ecosystem services and human well-being. The assessment deals with the full range of ecosystems—from those relatively undisturbed, such as natural forests, to landscapes with mixed patterns of human use and ecosystems intensively managed and modified by humans, such as agricultural land and urban areas.

A full assessment of the interactions between people and ecosystems requires a multiscale approach because it better reflects the multiscale nature of decision-making, allows the examination of driving forces that may be exogenous to particular regions, and provides a means of examining the differential impact of ecosystem changes and policy responses on different regions and groups within regions.

This section explains in greater detail the characteristics of each of the components of the MA conceptual framework, moving clockwise from the lower left corner of the figure in Box 2.

Ecosystems and Their Services

An ecosystem is a dynamic complex of plant, animal, and microorganism communities and the nonliving environment interacting as a functional unit. Humans are an integral part of ecosystems. Ecosystems provide a variety of benefits to people, including provisioning, regulating, cultural, and supporting services. Provisioning services are the products people obtain from ecosystems, such as food, fuel, fiber, fresh water, and genetic resources. Regulating services are the benefits people obtain from the regulation of ecosystem processes, including air quality maintenance, climate regulation, erosion control, regulation of human diseases, and water purification. Cultural services are the nonmaterial benefits people obtain from ecosystems through spiritual enrichment, cognitive development, reflection, recreation, and aesthetic experiences. Supporting services are those that are necessary for the production of all other ecosystem services, such as primary production, production of oxygen, and soil formation.

Biodiversity and ecosystems are closely related concepts. Biodiversity is the variability among living organisms from all sources, including terrestrial, marine, and other aquatic ecosystems and the ecological com-

---

BOX 2.   **Millennium Ecosystem Assessment Conceptual Framework**

Changes in factors that indirectly affect ecosystems, such as population, technology, and lifestyle (upper right corner of figure), can lead to changes in factors directly affecting ecosystems, such as the catch of fisheries or the application of fertilizers to increase food production (lower right corner). The resulting changes in the ecosystem (lower left corner) cause the ecosystem services to change and thereby affect human well-being. These interactions can take place at more than one scale and can cross scales. For example, a global market may lead to regional loss of forest cover, which increases flood magnitude along a local stretch of a river. Similarly, the interactions can take place across different time scales. Actions can be taken either to respond to negative changes or to enhance positive changes at almost all points in this framework (black cross bars).

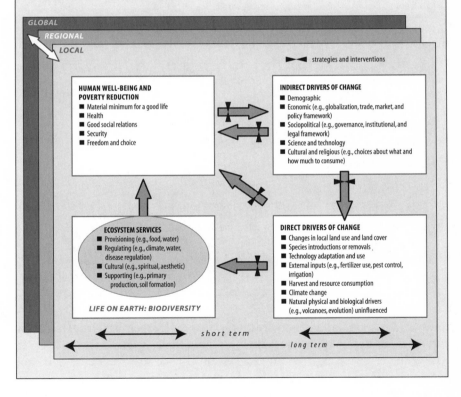

plexes of which they are part. It includes diversity within and between species and diversity of ecosystems. Diversity is a structural feature of ecosystems, and the variability among ecosystems is an element of biodiversity. Products of biodiversity include many of the services produced by ecosys-

BOX 3.   Reporting Categories Used in the Millennium Ecosystem Assessment

The MA will use 10 categories of systems to report its global findings. (See table.) These categories are not ecosystems themselves; each contains a number of ecosystems. The MA reporting categories are not mutually exclusive: their boundaries can and do overlap. Ecosystems within each category share a suite of biological, climatic, and social factors that tend to differ across categories. Because the boundaries of these reporting categories overlap, any place on Earth may fall into more than one category. Thus, for example, a wetland ecosystem in a coastal region may be examined both in the MA analysis of "coastal systems" as well as in its analysis of "inland water systems."

Millennium Ecosystem Assessment Reporting Categories

Category	Central Concept	Boundary Limits for Mapping
Marine	Ocean, with fishing typically a major driver of change	Marine areas where the sea is deeper than 50 meters.
Coastal	Interface between ocean and land, extending seawards to about the middle of the continental shelf and inland to include all areas strongly influenced by the proximity to the ocean	Area between 50 meters below mean sea level and 50 meters above the high tide level or extending landward to a distance 100 kilometers from shore. Includes coral reefs, intertidal zones, estuaries, coastal aquaculture, and seagrass communities.
Inland water	Permanent water bodies inland from the coastal zone, and areas whose ecology and use are dominated by the permanent, seasonal, or intermittent occurrence of flooded conditions	Rivers, lakes, floodplains, reservoirs, and wetlands; includes inland saline systems. Note that the Ramsar Convention considers "wetlands" to include both inland water and coastal categories.
Forest	Lands dominated by trees; often used for timber, fuelwood, and non-timber forest products	A canopy cover of at least 40 percent by woody plants taller than 5 meters. The existence of many other definitions is acknowledged, and other limits (such as crown cover greater than 10 percent, as used by the Food and Agriculture Organization of the United Nations) will also be reported. Includes temporarily cut-over forests and plantations; excludes orchards and agroforests where the main products are food crops.

tems (such as food and genetic resources), and changes in biodiversity can influence all the other services they provide. In addition to the important role of biodiversity in providing ecosystem services, the diversity of living species has intrinsic value independent of any human concern.

The concept of an ecosystem provides a valuable framework for analyzing and acting on the linkages between people and the environment. For that reason, the "ecosystem approach" has been endorsed by the Convention on Biological Diversity (CBD), and the MA conceptual framework is entirely consistent with this approach. The CBD states that the ecosys-

BOX 3.   continued

Millennium Ecosystem Assessment Reporting Categories

Category	Central Concept	Boundary Limits for Mapping
Dryland	Lands where plant production is limited by water availability; the dominant uses are large mammal herbivory, including livestock grazing, and cultivation	Drylands as defined by the Convention to Combat Desertification, namely lands where annual precipitation is less than two thirds of potential evaporation, from dry subhumid areas (ratio ranges 0.50–0.65), through semiarid, arid, and hyper-arid (ratio <0.05), but excluding polar areas; drylands include cultivated lands, scrublands, shrublands, grasslands, semi-deserts, and true deserts.
Island	Lands isolated by surrounding water, with a high proportion of coast to hinterland	As defined by the Alliance of Small Island States
Mountain	Steep and high lands	As defined by Mountain Watch using criteria based on elevation alone, and at lower elevation, on a combination of elevation, slope, and local elevation range. Specifically, elevation >2,500 meters, elevation 1,500–2,500 meters and slope >2 degrees, elevation 1,000–1,500 meters and slope >5 degrees or local elevation range (7 kilometers radius) >300 meters, elevation 300–1,000 meters and local elevation range (7 kilometers radius) >300 meters, isolated inner basins and plateaus less than 25 square kilometers extent that are surrounded by mountains.
Polar	High-latitude systems frozen for most of the year	Includes ice caps, areas underlain by permafrost, tundra, polar deserts, and polar coastal areas. Excludes high-altitude cold systems in low latitudes.
Cultivated	Lands dominated by domesticated plant species, used for and substantially changed by crop, agroforestry, or aquaculture production	Areas in which at least 30 percent of the landscape comes under cultivation in any particular year. Includes orchards, agroforestry, and integrated agriculture-aquaculture systems.
Urban	Built environments with a high human density	Known human settlements with a population of 5,000 or more, with boundaries delineated by observing persistent night-time lights or by inferring areal extent in the cases where such observations are absent.

tem approach is a strategy for the integrated management of land, water, and living resources that promotes conservation and sustainable use in an equitable way. This approach recognizes that humans, with their cultural diversity, are an integral component of many ecosystems.

In order to implement the ecosystem approach, decision-makers need to understand the multiple effects on an ecosystem of any management or policy change. By way of analogy, decision-makers would not take a decision about financial policy in a country without examining the condition of the economic system, since information on the economy of a single sector such as manufacturing would be insufficient. The same need to examine the consequences of changes for multiple sectors applies to ecosystems. For instance, subsidies for fertilizer use may increase food production, but sound decisions also require information on whether the potential reduction in the harvests of downstream fisheries as a result of water quality degradation from the fertilizer runoff might outweigh those benefits.

For the purpose of analysis and assessment, a pragmatic view of ecosystem boundaries must be adopted, depending on the questions being asked. A well-defined ecosystem has strong interactions among its components and weak interactions across its boundaries. A useful choice of ecosystem boundary is one where a number of discontinuities coincide, such as in the distribution of organisms, soil types, drainage basins, and depth in a waterbody. At a larger scale, regional and even globally distributed ecosystems can be evaluated based on a commonality of basic structural units. The global assessment being undertaken by the MA will report on marine, coastal, inland water, forest, dryland, island, mountain, polar, cultivated, and urban regions. These regions are not ecosystems themselves, but each contains a number of ecosystems (See Box 3.)

People seek multiple services from ecosystems and thus perceive the condition of given ecosystems in relation to their ability to provide the services desired. Various methods can be used to assess the ability of ecosystems to deliver particular services. With those answers in hand, stakeholders have the information they need to decide on a mix of services best meeting their needs. The MA will consider criteria and methods to provide an integrated view of the condition of ecosystems. The condition of each category of ecosystem services is evaluated in somewhat different ways, although in general a full assessment of any service requires considerations of stocks, flows, and resilience of the service.

Human Well-being and Poverty Reduction

Human well-being has multiple constituents, including the basic material for a good life, freedom and choice, health, good social relations, and security. Poverty is also multidimensional and has been defined as the pronounced deprivation of well-being. How well-being, ill-being, or poverty are experienced and expressed depends on context and situation, reflecting local physical, social, and personal factors such as geography, environment, age, gender, and culture. In all contexts, however, ecosystems are essential for human well-being through their provisioning, regulating, cultural, and supporting services.

Human intervention in ecosystems can amplify the benefits to human society. However, evidence in recent decades of escalating human impacts on ecological systems worldwide raises concerns about the spatial and temporal consequences of ecosystem changes detrimental to human well-being. Ecosystem changes affect human well-being in the following ways:

- **Security** is affected both by changes in provisioning services, which affect supplies of food and other goods and the likelihood of conflict over declining resources, and by changes in regulating services, which could influence the frequency and magnitude of floods, droughts, landslides, or other catastrophes. It can also be affected by changes in cultural services as, for example, when the loss of important ceremonial or spiritual attributes of ecosystems contributes to the weakening of social relations in a community. These changes in turn affect material well-being, health, freedom and choice, security, and good social relations.

- **Access to basic material for a good life** is strongly linked to both provisioning services such as food and fiber production and regulating services, including water purification.

- **Health** is strongly linked to both provisioning services such as food production and regulating services, including those that influence the distribution of disease-transmitting insects and of irritants and pathogens in water and air. Health can also be linked to cultural services through recreational and spiritual benefits.

- **Social relations** are affected by changes to cultural services, which affect the quality of human experience.

- **Freedoms and choice** are largely predicated on the existence of the other components of well-being and are thus influenced by changes in provisioning, regulating, or cultural services from ecosystems.

Human well-being can be enhanced through sustainable human interactions with ecosystems supported by necessary instruments, institutions, organizations, and technology. Creation of these through participation and transparency may contribute to freedoms and choice as well as to increased economic, social, and ecological security. By ecological security, we mean the minimum level of ecological stock needed to ensure a sustainable flow of ecosystem services.

Yet the benefits conferred by institutions and technology are neither automatic nor equally shared. In particular, such opportunities are more readily grasped by richer than poorer countries and people; some institutions and technologies mask or exacerbate environmental problems; responsible governance, while essential, is not easily achieved; participation in decision-making, an essential element of responsible governance, is expensive in time and resources to maintain. Unequal access to ecosystem services has often elevated the well-being of small segments of the population at the expense of others.

Sometimes the consequences of the depletion and degradation of ecosystem services can be mitigated by the substitution of knowledge and of manufactured or human capital. For example, the addition of fertilizer in agricultural systems has been able to offset declining soil fertility in many regions of the world where people have sufficient economic resources to purchase these inputs, and water treatment facilities can sometimes substitute for the role of watersheds and wetlands in water purification. But ecosystems are complex and dynamic systems and there are limits to substitution possibilities, especially with regulating, cultural, and supporting services. No substitution is possible for the extinction of culturally important species such as tigers or whales, for instance, and substitutions may be economically impractical for the loss of services such as erosion control or climate regulation. Moreover, the scope for substitutions varies by social, economic, and cultural conditions. For some people, especially the poorest, substitutes and choices are very limited. For those who are better off, substitution may be possible through trade, investment, and technology.

Because of the inertia in both ecological and human systems, the consequences of ecosystem changes made today may not be felt for decades. Thus, sustaining ecosystem services, and thereby human well-being, requires a full understanding and wise management of the relationships between human activities, ecosystem change, and well-being over the short, medium, and long term. Excessive current use of ecosystem services compromises their future availability. This can be prevented by ensuring that the use is sustainable.

Achieving sustainable use requires effective and efficient institutions that can provide the mechanisms through which concepts of freedom, justice, fairness, basic capabilities, and equity govern the access to and use of ecosystem services. Such institutions may also need to mediate conflicts between individual and social interests that arise.

The best way to manage ecosystems to enhance human well-being will differ if the focus is on meeting needs of the poor and weak or the rich and powerful. For both groups, ensuring the long-term supply of ecosystem services is essential. But for the poor, an equally critical need is to provide more equitable and secure access to ecosystem services.

Drivers of Change

Understanding the factors that cause changes in ecosystems and ecosystem services is essential to designing interventions that capture positive impacts and minimize negative ones. In the MA, a "driver" is any factor that changes an aspect of an ecosystem. A direct driver unequivocally influences ecosystem processes and can therefore be identified and measured to differing degrees of accuracy. An indirect driver operates more diffusely, often by altering one or more direct drivers, and its influence is established by understanding its effect on a direct driver. Both indirect and direct drivers often operate synergistically. Changes in land cover, for example, can increase the likelihood of introduction of alien invasive species. Similarly, technological advances can increase rates of economic growth.

The MA explicitly recognizes the role of decision-makers who affect ecosystems, ecosystem services, and human well-being. Decisions are made at three organizational levels, although the distinction between those levels is often diffuse and difficult to define:

- by individuals and small groups at the local level (such as a field or forest stand) who directly alter some part of the ecosystem;
- by public and private decision-makers at the municipal, provincial, and national levels; and
- by public and private decision-makers at the international level, such as through international conventions and multilateral agreements.

The decision-making process is complex and multidimensional. We refer to a driver that can be influenced by a decision-maker as an endogenous driver and one over which the decision-maker does not have control as an exogenous driver. The amount of fertilizer applied on a farm is an endogenous driver from the standpoint of the farmer, for example, while the price of the fertilizer is an exogenous driver, since the farmer's decisions have little direct influence on price. The specific temporal, spatial, and organizational scale dependencies of endogenous and exogenous drivers and the specific linkages and interactions among drivers will be explicitly assessed in the MA.

Whether a driver is exogenous or endogenous to a decision-maker is dependent upon the spatial and temporal scale. For example, a local decision-maker can directly influence the choice of technology, changes in land use, and external inputs (such as fertilizers or irrigation), but has little control over prices and markets, property rights, technology development, or the local climate. In contrast, a national or regional decision-maker has more control over many factors, such as macroeconomic policy, technology development, property rights, trade barriers, prices, and markets. But on the short time scale, that individual has little control over the climate or global population. On the longer time scale, drivers that are exogenous to a decision-maker in the short run, such as population, become endogenous since the decision-maker can influence them through, for instance, education, the advancement of women, and migration policies.

The indirect drivers of change are primarily:

- demographic (such as population size, age and gender structure, and spatial distribution);

- economic (such as national and per capita income, macroeconomic policies, international trade, and capital flows);

- sociopolitical (such as democratization, the roles of women, of civil society, and of the private sector, and international dispute mechanisms);

- scientific and technological (such as rates of investments in research and development and the rates of adoption of new technologies, including biotechnologies and information technologies); and

- cultural and religious (such as choices individuals make about what and how much to consume and what they value).

The interaction of several of these drivers, in turn, affects levels of resource consumption and differences in consumption both within and between countries. Clearly these drivers are changing—population and the world economy are growing, for instance, there are major advances in information technology and biotechnology, and the world is becoming more interconnected. Changes in these drivers are projected to increase the demand for and consumption of food, fiber, clean water, and energy, which will in turn affect the direct drivers. The direct drivers are primarily physical, chemical, and biological—such as land cover change, climate change, air and water pollution, irrigation, use of fertilizers, harvesting, and the introduction of alien invasive species. Change is apparent here too: the climate is changing, species ranges are shifting, alien species are spreading, and land degradation continues.

An important point is that any decision can have consequences external to the decision framework. These consequences are called externalities because they are not part of the decision-making calculus. Externalities can have positive or negative effects. For example, a decision to subsidize fertilizers to increase crop production might result in substantial degradation of water quality from the added nutrients and degradation of downstream fisheries. But it is also possible to have positive externalities. A beekeeper might be motivated by the profits to be made from selling honey, for instance, but neighboring orchards could produce more apples because of enhanced pollination arising from the presence of the bees.

Multiple interacting drivers cause changes in ecosystem services. There are functional interdependencies between and among the indirect and direct drivers of change, and, in turn, changes in ecological services lead to feedbacks on the drivers of changes in ecological services. Synergetic driver combinations are common. The many processes of globalization lead to new forms of interactions between drivers of changes in ecosystem services.

Cross-scale Interactions and Assessment

An effective assessment of ecosystems and human well-being cannot be conducted at a single temporal or spatial scale. Thus the MA conceptual framework includes both of these dimensions. Ecosystem changes that may have little impact on human well-being over days or weeks (soil erosion, for instance) may have pronounced impacts over years or decades (declining agricultural productivity). Similarly, changes at a local scale may have little impact on some services at that scale (as in the local impact of forest loss on water availability) but major impacts at large scales (forest loss in a river basin changing the timing and magnitude of downstream flooding).

Ecosystem processes and services are typically most strongly expressed, are most easily observed, or have their dominant controls or consequences at particular spatial and temporal scales. They often exhibit a characteristic scale—the typical extent or duration over which processes have their impact. Spatial and temporal scales are often closely related. For instance, food production is a localized service of an ecosystem and changes on a weekly basis, water regulation is regional and changes on a monthly or seasonal basis, and climate regulation may take place at a global scale over decades.

Assessments need to be conducted at spatial and temporal scales appropriate to the process or phenomenon being examined. Those done over large areas generally use data at coarse resolutions, which may not detect fine-resolution processes. Even if data are collected at a fine level of detail, the process of averaging in order to present findings at the larger scale causes local patterns or anomalies to disappear. This is particularly problematic for processes exhibiting thresholds and nonlinearities. For example, even though a number of fish stocks exploited in a particular area might have collapsed due to overfishing, average catches across all stocks (including healthier stocks) would not reveal the extent of the problem. Assessors, if they are aware of such thresholds and have access to high-resolution data, can incorporate such information even in a large-scale assessment. Yet an assessment done at smaller spatial scales can help identify important dynamics of the system that might otherwise be overlooked. Likewise, phenomena and processes that occur at much larger scales, although expressed locally, may go unnoticed in purely local-scale assessments. Increased carbon dioxide concentrations or decreased stratospheric ozone concentrations have local effects, for instance, but it would be difficult to trace the causality of the effects without an examination of the overall global process.

Time scale is also very important in conducting assessments. Humans tend not to think beyond one or two generations. If an assessment covers a shorter time period than the characteristic temporal scale, it may not adequately capture variability associated with long-term cycles, such as glaciation. Slow changes are often harder to measure, as is the case with the impact of climate change on the geographic distribution of species or populations. Moreover, both ecological and human systems have substantial inertia, and the impact of changes occurring today may not be seen for years or decades. For example, some fisheries catches may increase for several years even after they have reached unsustainable levels because of the large number of juvenile fish produced before that level was reached.

Social, political, and economic processes also have characteristic scales, which may vary widely in duration and extent. Those of ecological and sociopolitical processes often do not match. Many environmental problems originate from this mismatch between the scale at which the ecological process occurs, the scale at which decisions are made, and the scale of institutions for decision-making. A purely local-scale assessment, for instance, may discover that the most effective societal response requires action that can occur only at a national scale (such as the removal of a subsidy or the establishment of a regulation). Moreover, it may lack the relevance and credibility necessary to stimulate and inform national or regional changes. On the other hand, a purely global assessment may lack both the relevance and the credibility necessary to lead to changes in ecosystem management at the local scale where action is needed. Outcomes at a given scale are often heavily influenced by interactions of ecological, socioeconomic, and political factors emanating from other scales. Thus focusing solely on a single scale is likely to miss interactions with other scales that are critically important in understanding ecosystem determinants and their implications for human well-being.

The choice of the spatial or temporal scale for an assessment is politically laden, since it may intentionally or unintentionally privilege certain groups. The selection of assessment scale with its associated level of detail implicitly favors particular systems of knowledge, types of information, and modes of expression over others. For example, non-codified information or knowledge systems of minority populations are often missed when assessments are undertaken at larger spatial scales or higher levels of aggregation. Reflecting on the political consequences of scale and boundary choices is an important prerequisite to exploring what multi- and cross-scale analysis in the MA might contribute to decision-making and public policy processes at various scales.

Values Associated with Ecosystems

Current decision-making processes often ignore or underestimate the value of ecosystem services. Decision-making concerning ecosystems and their services can be particularly challenging because different disciplines, philosophical views, and schools of thought assess the value of ecosystems differently. One paradigm of value, known as the utilitarian (anthropocentric) concept, is based on the principle of humans' preference satisfaction (welfare). In this case, ecosystems and the services they provide have value to human societies because people derive utility from their use, either directly or indirectly (use values). Within this utilitarian concept of value, people also give value to ecosystem services that they are not currently using (non-use values). Non-use values, usually known as existence value, involve the case where humans ascribe value to knowing that a resource exists even if they never use that resource directly. These often involve the deeply held historical, national, ethical, religious, and spiritual values people ascribe to ecosystems—the values that the MA recognizes as cultural services of ecosystems.

A different, non-utilitarian value paradigm holds that something can have intrinsic value—that is, it can be of value in and for itself—irrespective of its utility for someone else. From the perspective of many ethical, religious, and cultural points of view, ecosystems may have intrinsic value, independent of their contribution to human well-being.

The utilitarian and non-utilitarian value paradigms overlap and interact in many ways, but they use different metrics, with no common denominator, and cannot usually be aggregated, although both paradigms of value are used in decision-making processes.

Under the utilitarian approach, a wide range of methodologies has been developed to attempt to quantify the benefits of different ecosystem services. These methods are particularly well developed for provisioning services, but recent work has also improved the ability to value regulating and other services. The choice of valuation technique in any given instance is dictated by the characteristics of the case and by data availability. (See Box 4.)

Non-utilitarian value proceeds from a variety of ethical, cultural, religious, and philosophical bases. These differ in the specific entities that are deemed to have intrinsic value and in the interpretation of what having intrinsic value means. Intrinsic value may complement or counterbalance considerations of utilitarian value. For example, if the aggregate utility of the services provided by an ecosystem (as measured by its utilitarian value) outweighs the value of converting it to another use, its intrinsic value may then be complementary and provide an additional impetus for conserving the ecosystem. If, however, economic valuation indicates that the value of converting the ecosystem outweighs the aggregate value of its services, its ascribed intrinsic value may be deemed great enough to warrant a social decision to conserve it anyway. Such decisions are essentially political, not economic. In contemporary democracies these decisions are made by parliaments or legislatures or by regulatory agencies mandated to do so by law. The sanctions for violating laws recognizing an entity's intrinsic value may be regarded as a measure of the degree of intrinsic value ascribed to them. The decisions taken by businesses, local communities, and individuals also can involve considerations of both utilitarian and non-utilitarian values.

The mere act of quantifying the value of ecosystem services cannot by itself change the incentives affecting their use or misuse. Several changes in current practice may be required to take better account of these values. The MA will assess the use of information on ecosystem service values in decision-making. The goal is to improve decision-making processes and tools and to provide feedback regarding the kinds of information that can have the most influence.

Assessment Tools

The information base exists in any country to undertake an assessment within the framework of the MA. That said, although new data sets (for

---

BOX 4.   Valuation of Ecosystem Services

Valuation can be used in many ways: to assess the total contribution that ecosystems make to human well-being, to understand the incentives that individual decision-makers face in managing ecosystems in different ways, and to evaluate the consequences of alternative courses of action. The MA plans to use valuation primarily in the latter sense: as a tool that enhances the ability of decision-makers to evaluate trade-offs between alternative ecosystem management regimes and courses of social actions that alter the use of ecosystems and the multiple services they provide. This usually requires assessing the change in the mix (the value) of services provided by an ecosystem resulting from a given change in its management.

Most of the work involved in estimating the change in the value of the flow of benefits provided by an ecosystem involves estimating the change in the physical flow of benefits (quantifying biophysical relations) and tracing through and quantifying a chain of causality between changes in ecosystem condition and human welfare. A common problem in valuation is that information is only available on some of the links in the chain and often in incompatible units. The MA can make a major contribution by making various disciplines better aware of what is needed to ensure that their work can be combined with that of others to allow a full assessment of the consequences of altering ecosystem state and function.

The ecosystem values in this sense are only one of the bases on which decisions on ecosystem management are and should be made. Many other factors, including notions of intrinsic value and other objectives that society might have (such as equity among different groups or generations), will also feed into the decision framework. Even when decisions are made on other bases, however, estimates of changes in utilitarian value provide invaluable information.

---

example, from remote sensing) providing globally consistent information make a global assessment like the MA more rigorous, there are still many challenges that must be dealt with in using these data at global or local scales. Among these challenges are biases in the geographic and temporal coverage of the data and in the types of data collected. Data availability for industrial countries is greater than that for developing ones, and data for certain resources such as crop production are more readily available than data for fisheries, fuelwood, or biodiversity. The MA makes extensive use of both biophysical and socioeconomic indicators, which combine data into policy-relevant measures that provide the basis for assessment and decision-making.

Models can be used to illuminate interactions among systems and drivers, as well as to make up for data deficiencies—for instance, by providing estimates where observations are lacking. The MA will make use of environmental system models that can be used, for example, to measure the consequences of land cover change for river flow or the consequences of climate change for the distribution of species. It will also use human system models that can examine, for instance, the impact of changes in ecosystems on production, consumption, and investment decisions by households or that allow the economy-wide impacts of a change in production in a particular sector like agriculture to be evaluated. Finally, integrated models, combining both the environmental and human systems linkages, can increasingly be used at both global and sub-global scales.

The MA aims to incorporate both formal scientific information and traditional or local knowledge. Traditional societies have nurtured and refined systems of knowledge of direct value to those societies but also of considerable value to assessments undertaken at regional and global scales. This information often is unknown to science and can be an expression of other relationships between society and nature in general and of sustainable ways of managing natural resources in particular. To be credible and useful to decision-makers, all sources of information, whether scientific, traditional, or practitioner knowledge, must be critically assessed and validated as part of the assessment process through procedures relevant to the form of knowledge.

Since policies for dealing with the deterioration of ecosystem services are concerned with the future consequences of current actions, the development of scenarios of medium- to long-term changes in ecosystems, services, and drivers can be particularly helpful for decision-makers. Scenarios are typically developed through the joint involvement of decision-makers and scientific experts, and they represent a promising mechanism for linking scientific information to decision-making processes. They do not attempt to predict the future but instead are designed to indicate what science can and cannot say about the future consequences of alternative plausible choices that might be taken in the coming years.

The MA will use scenarios to summarize and communicate the diverse trajectories that the world's ecosystems may take in future decades. Scenarios are plausible alternative futures, each an example of what might happen under particular assumptions. They can be used as a systematic method for thinking creatively about complex, uncertain futures. In this way, they help us understand the upcoming choices that need to be made and highlight developments in the present. The MA will develop scenarios that connect possible changes in drivers (which may be unpredictable or uncontrollable) with human demands for ecosystem services. The scenarios will link these demands, in turn, to the futures of the services themselves and the aspects of human welfare that depend on them. The scenario building exercise will break new ground in several areas:

- development of scenarios for global futures linked explicitly to ecosystem services and the human consequences of ecosystem change,

- consideration of trade-offs among individual ecosystem services within the "bundle" of benefits that any particular ecosystem potentially provides to society,

- assessment of modeling capabilities for linking socioeconomic drivers and ecosystem services, and

- consideration of ambiguous futures as well as quantifiable uncertainties.

The credibility of assessments is closely linked to how they address what is not known in addition to what is known. The consistent treatment of uncertainty is therefore essential for the clarity and utility of assessment reports. As part of any assessment process, it is crucial to estimate the uncertainty of findings even if a detailed quantitative appraisal of uncertainty is unavailable.

Strategies and Interventions

The MA will assess the use and effectiveness of a wide range of options for responding to the need to sustainably use, conserve, and restore ecosystems and the services they provide. These options include incorporating the value of ecosystems in decisions, channeling diffuse ecosystem benefits to decision-makers with focused local interests, creating markets and property rights, educating and dispersing knowledge, and investing to improve ecosystems and the services they provide. As seen in Box 2 on the MA conceptual framework, different types of response options can affect the relationships of indirect to direct drivers, the influence of direct drivers on ecosystems, the human demand for ecosystem services, or the impact of changes in human well-being on indirect drivers. An effective strategy for managing ecosystems will involve a mix of interventions at all points in this conceptual framework.

Mechanisms for accomplishing these interventions include laws, regulations, and enforcement schemes; partnerships and collaborations; the sharing of information and knowledge; and public and private action. The choice of options to be considered will be greatly influenced by both the temporal and the physical scale influenced by decisions, the uncertainty of outcomes, cultural context, and the implications for equity and trade-offs. Institutions at different levels have different response options available to them, and special care is required to ensure policy coherence.

Decision-making processes are value-based and combine political and technical elements to varying degrees. Where technical input can play a role, a range of tools is available to help decision-makers choose among strategies and interventions, including cost-benefit analysis, game theory, and policy exercises. The selection of analytical tools should be determined by the context of the decision, key characteristics of the decision problem, and the criteria considered to be important by the decision-makers. Information from these analytical frameworks is always combined with the intuition, experience, and interests of the decision-maker in shaping the final decisions.

Risk assessment, including ecological risk assessment, is an established discipline and has a significant potential for informing the decision process. Finding thresholds and identifying the potential for irreversible change are important for the decision-making process. Similarly, environmental impact assessments designed to evaluate the impact of particular projects and strategic environmental assessments designed to evaluate the impact of policies both represent important mechanisms for incorporating the findings of an ecosystem assessment into decision-making processes.

Changes also may be required in decision-making processes themselves. Experience to date suggests that a number of mechanisms can improve the process of making decisions about ecosystem services. Broadly accepted norms for decision-making process include the following characteristics. Did the process:

- bring the best available information to bear?
- function transparently, use locally grounded knowledge, and involve all those with an interest in a decision?
- pay special attention to equity and to the most vulnerable populations?
- use decision analytical frameworks that take account of the strengths and limits of individual, group, and organizational information processing and action?
- consider whether an intervention or its outcome is irreversible and incorporate procedures to evaluate the outcomes of actions and learn from them?
- ensure that those making the decisions are accountable?
- strive for efficiency in choosing among interventions?

■ take account of thresholds, irreversibility, and cumulative, cross-scale, and marginal effects and of local, regional, and global costs, risk, and benefits?

The policy or management changes made to address problems and opportunities related to ecosystems and their services, whether at local scales or national or international scales, need to be adaptive and flexible in order to benefit from past experience, to hedge against risk, and to consider uncertainty. The understanding of ecosystem dynamics will always be limited, socioeconomic systems will continue to change, and outside determinants can never be fully anticipated. Decision-makers should consider whether a course of action is reversible and should incorporate, whenever possible, procedures to evaluate the outcomes of actions and learn from them. Debate about exactly how to do this continues in discussions of adaptive management, social learning, safe minimum standards, and the precautionary principle. But the core message of all approaches is the same: acknowledge the limits of human understanding, give special consideration to irreversible changes, and evaluate the impacts of decisions as they unfold.

1 Introduction and Conceptual Framework

EXECUTIVE SUMMARY

- The goal of the Millennium Ecosystem Assessment (MA) is to establish the scientific basis for actions needed to enhance the contribution of ecosystems to human well-being without undermining their long-term productivity.

- The conceptual framework for the MA places human well-being as the central focus for assessment while recognizing that biodiversity and ecosystems also have intrinsic value and that people take decisions concerning ecosystems based on considerations of both well-being and intrinsic value.

- The MA conceptual framework assumes that a dynamic interaction exists between people and ecosystems, with the changing human condition serving to both directly and indirectly drive change in ecosystems and with changes in ecosystems causing changes in human well-being. At the same time, many other factors independent of the environment change the human condition, and many natural forces influence ecosystems.

- A full assessment of the interactions between people and ecosystems requires a multiscale approach, as this better reflects the multiscale nature of decision-making, allows the examination of driving forces from outside particular regions, and provides a means of examining the differential impact of ecosystem changes and policy responses on different regions and groups within regions.

- Effective incorporation of different types of knowledge in an assessment can both improve the findings and help to increase their adoption by stakeholders if they see that their information has contributed to those findings.

- The usefulness of an assessment can be enhanced by identifying and seeking to address its structural biases. Any assessment empowers some stakeholders at the expense of others by virtue of the selection of issues and of expert knowledge to be incorporated.

Introduction

Human well-being and progress toward sustainable development are vitally dependent upon Earth's ecosystems. The ways in which ecosystems are affected by human activities will have consequences for the supply of ecosystem services—including food, fresh water, fuelwood, and fiber—and for the prevalence of diseases, the frequency and magnitude of floods and droughts, and local as well as global climate. Ecosystems also provide spiritual, recreational, educational, and other nonmaterial benefits to people. Changes in availability of all these ecosystem services can profoundly affect aspects of human well-being—ranging from the rate of economic growth and health and livelihood security to the prevalence and persistence of poverty.

Human demands for ecosystem services are growing rapidly. At the same time, humans are altering the capability of ecosystems to continue to provide many of these services. Management of this relationship is required to enhance the contribution of ecosystems to human well-being without affecting their long-term capacity to provide services. The Millennium Ecosystem Assessment (MA) was established in 2001 by a partnership of international institutions, and with support from governments, with the goal of enhancing the scientific basis for such management.

The MA is being carried out through four working groups on condition and trends, scenarios, responses, and sub-global assessments. Each working group will involve a regionally balanced group of between 50 and 400 experts from dozens of countries as authors. The MA was launched in June 2001, the full assessment reports will undergo two rounds of peer-review by governments and experts in 2004, and the assessment reports will be released in 2005. Five short synthesis reports containing the key policy-relevant findings will also be released at that time focused on the needs of particular users such as the international conventions and the private sector. The MA includes at this time up to 15 sub-global assessments that are applying the MA conceptual framework and methodology to assessments designed to meet needs at local, national, and regional scales, and the products of these assessments will be released over the next three years. Throughout the MA process, an ongoing dialogue is taking place involving experts preparing the assessment and intended users of the findings in order to focus the assessment on the needs of users and to ensure that users are sufficiently engaged in the process that they will be able to make direct use of the findings.

All economies depend on ecosystem services. The production and manufacture of industrial wood products in the early 1990s contributed on the order of $400 billion to the global economy (Matthews et al. 2000). The world's fisheries contributed $55 billion in export value in 2000 (FAO 2000). Ecosystem services are particularly important to the economies of low-income developing countries. Between 1996 and 1998, for example, agriculture represented nearly one fourth of the total gross domestic product of low-income countries (Wood et al. 2000).

Certain ecosystem services—such as inland fisheries and fuelwood production—are particularly important to the livelihoods of poor people. Fisheries provide the primary source of animal protein for nearly 1 billion people, and all but 4 of the 30 countries most dependent on fish as a protein source are in the developing world (WRI et al. 2000). In Cambodia, for instance, roughly 60 percent of the total animal protein consumed is from the fishery resources of the Tonle Sap, a large freshwater lake. In Malawi, freshwater fisheries supply 70–75 percent of the animal protein for both urban and rural low-income families (WRI et al. 2000). Similarly, more than 2 billion people depend directly on biomass fuels as their primary or sole source of energy, and in countries like Nepal, Uganda, Rwanda, and Tanzania, woodfuel meets 80 percent or more of total energy requirements (Matthews et al. 2000). Moreover, poor people are highly vulnerable to health risks associated with ecosystems: some 1–3 million people die each year from malaria, with 90 percent of them in Africa, where problems of poverty are most pressing (WHO 1997).

Yet many ecosystem services are largely unrecognized in their global importance or in the pivotal role they play in meeting needs in particular countries and regions (Daily 1997a). For example, terrestrial and ocean ecosystems provide a tremendous service by absorbing nearly 60 percent of the carbon that is now emitted to the atmosphere from human activities (IPCC 2000), thereby slowing the rate of global climate change. A number of cities—including New York and Portland, Oregon, in the United States, Caracas in Venezuela, and Curitiba in Brazil—reduce water treatment costs by investing in the protection of the natural water quality regulation provided by well-managed ecosystems (Reid 2001). The contribution of pollination to the worldwide production of 30 major fruit, vegetable, and tree crops is estimated to be approximately $54 billion a year (Kenmore and Krell 1998). Even in urban centers, ecosystems contribute significantly to well-being, both aesthetically and economically: Chicago's trees remove more than 5,000 tons of pollutants a year from the atmosphere (Nowak 1994).

A society's "natural capital"—its living and nonliving resources—is a key determinant of its well-being. The full wealth of a nation can be evaluated only with due consideration to all forms of capital: manufactured, human, social, and natural. (See Figure 1.1.) Historically, given the abundant supply of natural capital and the application of new technologies to enhance the production of certain services, humanity has been remark-

FIGURE 1.1   Society's Productive Base

A society's productive base is composed of four types of capital: manufactured, human, social, and natural.

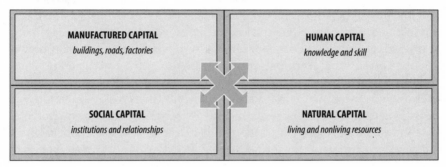

MANUFACTURED CAPITAL	HUMAN CAPITAL
buildings, roads, factories	*knowledge and skill*
SOCIAL CAPITAL	NATURAL CAPITAL
institutions and relationships	*living and nonliving resources*

ably successful in meeting growing demands for particular services. Between 1967 and 1982, for example, conversion of native ecosystems to agricultural ecosystems, combined with a 2.2-percent annual increase in cereal yields, led to net increases in per capita food availability even though there was simultaneously a 32-percent increase in world population (Pinstrup-Andersen et al. 1997). But despite the success in meeting growth in aggregate demand, there have been significant problems in meeting demands in particular regions. Moreover, increased supply of certain goods, such as food, has often meant a trade-off with the supply of other ecosystem services, such as protecting water quality or supplying timber.

Current demands for ecosystem services are growing rapidly and often already outstrip capacity. Between 1993 and 2020, world demand for rice, wheat, and maize is projected to increase by some 40 percent and livestock production by more than 60 percent (Pinstrup-Andersen et al. 1997). Humans now withdraw about 20 percent of the base flow of the world's rivers, and during the past century withdrawals grew twice as fast as world population (Shiklomanov 1997; WHO 1997). By 2020, world use of industrial roundwood could be anywhere from 23 to 55 percent over 1998 consumption levels (Brooks et al. 1996).

These growing demands can no longer be met by tapping unexploited resources (Watson et al. 1998; Ayensu et al. 2000). A country can increase food supply by converting a forest to agriculture, but in so doing it decreases the supply of goods that may be of equal or greater importance, such as clean water, timber, biodiversity, or flood control. Even more significant, humans are increasingly undermining the productive capability of ecosystems to provide the services that people desire. For example, world fisheries are now declining due to overfishing, and some 40 percent of agricultural land has been strongly or very strongly degraded in the past 50 years by erosion, salinization, compaction, nutrient depletion, biological degradation, or pollution (WRI et al. 2000).

Continuing degradation of the world's ecosystems is neither inevitable nor justified. Many instruments now exist that can aid in the management of human demand for ecosystem services and of impacts of human activities on ecosystems. Recent progress in cost-effective technologies, policies, and regulation can contribute to management systems that can reduce and eventually reverse many of today's problems. Investments in improved management of ecosystem services tend to be highly leveraged strategies for sustainable development. Like the benefits of increased education or improved governance, the protection, restoration, and enhancement of ecosystem services tend to have multiple and synergistic benefits. For example, technology allows partial substitution of the ecosystem service of water purification through the construction of water treatment facilities. But by protecting the watershed to enable the ecosystem to provide this service instead, a variety of other benefits can often be obtained—such as the maintenance of fisheries, reduction of flood risks, and protection of recreational and amenity values.

New policies and initiatives in diverse economies and cultures illustrate practical mechanisms for protecting vital ecosystem services and enhancing their contributions to human development. More effective balances in the supply of various services can often be restored: reduction of subsidies that have contributed to excessive fishing harvest in many fisheries, for instance, can lessen harvest pressure now, protect biodiversity, and ultimately lead to increased catch per unit of effort.

Institutional arrangements such as changes in land tenure or rights to resources can help ensure that those paying to protect ecosystem services receive a fair share of the benefits: some power companies, for example, are now paying countries to protect and restore forests for their carbon sequestration service as a means of offsetting carbon emissions (Daily and Ellison 2002). And in Costa Rica, a new national program pays private landowners for a suite of ecosystem services flowing from forested (and reforested) land, including watershed protection, biodiversity conservation, and preservation of scenic beauty (Castro et al. 1998). Techniques for restoration can also be used: in the Murray-Darling River Basin of Australia, which supplies 75 percent of Australia's irrigation water and over 40 percent of the nation's agricultural production, native vegetation is being replanted as a cost-effective tool in con-

---

BOX 1.1   Commitment to Sustainable Development

The interlinkages among environmental management, poverty alleviation, and sustainable development have long been recognized by governments and international institutions. Examples of conferences, initiatives, and reports that have stressed this theme in recent years include:

Conferences and Initiatives
- United Nations Conference on the Human Environment (Stockholm, 1972)
- United Nations Conference on Environment and Development (Rio de Janeiro, 1992)
- World Conference on Human Rights (Vienna, 1993)
- International Conference on Population and Development (Cairo, 1994)
- Global Conference on the Sustainable Development of Small Island Developing States (Bridgetown, 1994)
- World Summit for Social Development (Copenhagen, 1995)
- World Conference on Women (Beijing, 1995)
- World Food Summit (Rome, 1996)
- United Nations Millennium Summit (New York, 2000)
- Initiative for the Heavily Indebted Poor Countries (2001)
- World Summit on Sustainable Development (Johannesburg, 2002)

Reports and Statements
- *World Conservation Strategy* (IUCN et al. 1980)
- *Our Common Future* (WCED 1987)
- *Caring for the Earth* (IUCN et al. 1991)
- *Statement on Population* (statement of 58 scientific academies, 1994)
- *The Challenges of an Urban World* (statement of 72 scientific academies, 1996)
- *Our Common Journey: A Transition Toward Sustainability* (NRC 1999)
- *United Nations Millennium Declaration* (2000)
- *Transition to Sustainability in the 21st Century: The Contribution of Science and Technology* (statement of 73 scientific academies, 2000)

---

trolling devastating salinization of farmland (Murray-Darling Basin Ministerial Council 2001).

Roughly half of the world's poorest people live in marginal areas such as arid lands, steep slopes, or coastal margins that are prone to degradation and highly vulnerable to floods, droughts, or landslides (UNDP 1998). Some 80 percent of poor people in developing countries live in rural areas where people directly harvest ecosystem goods (Jazairy et al. 1992). Ap-

---

BOX 1.2   Millennium Development Goals

The Millennium Development Goals were adopted in September 2000 during the 55th Session of the United Nations General Assembly, known as the Millennium Assembly.

Goal 1: Eradicate extreme poverty and hunger
- Halve, between 1990 and 2015, the proportion of people whose income is less than one dollar a day.
- Halve, between 1990 and 2015, the proportion of people who suffer from hunger.

Goal 2: Achieve universal primary education
- Ensure that, by 2015, children everywhere, boys and girls alike, will be able to complete a full course of primary schooling.

Goal 3: Promote gender equality and empower women
- Eliminate gender disparity in primary and secondary education preferably by 2005 and at all levels of education no later than 2015.

Goal 4: Reduce child mortality
- Reduce by two thirds, between 1990 and 2015, the under-five mortality rate.

Goal 5: Improve maternal health
- Reduce by three quarters, between 1990 and 2015, the maternal mortality ratio.

Goal 6: Combat HIV/AIDS, malaria, and other diseases
- Have halted by 2015, and begun to reverse, the spread of HIV/AIDS.
- Have halted by 2015, and begun to reverse, the incidence of malaria and other major diseases.

---

proaches to poverty alleviation through environmental management can provide cost-effective and lasting solutions that often work in concert with education, empowerment of women, and improved governance. Fortunately, the need for more effective investment in ecosystem management is increasingly being recognized by governments as a tool for poverty alleviation.

Various conferences and reports over the past two decades, culminating in the 2002 World Summit on Sustainable Development, have outlined key principles of a more socially responsible and environmentally sustainable world for both industrial and developing countries, recogniz-

---

BOX 1.2   continued

Goal 7: Ensure environmental sustainability

- Integrate the principles of sustainable development into country policies and programmes and reverse the loss of environmental resources.
- Halve, by 2015, the proportion of people without sustainable access to safe drinking water.
- By 2020, have achieved a significant improvement in the lives of at least 100 million slum dwellers.

Goal 8: Develop a global partnership for development

- Develop further an open, rule-based, predictable, non-discriminatory trading and financial system (includes a commitment to good governance, development, and poverty reduction—both nationally and internationally).
- Address the least developed countries' special needs (includes tariff- and quota-free access for their exports, enhanced debt relief for heavily indebted poor countries, cancellation of official bilateral debt, and more generous official development assistance for countries committed to poverty reduction).
- Address the special needs of landlocked countries and small island developing states (through Barbados Programme and 22nd General Assembly provisions).
- Deal comprehensively with the debt problems of developing countries through national and international measures in order to make debt sustainable in the long term.
- In cooperation with developing countries, develop and implement strategies for decent and productive work for youth.
- In cooperation with pharmaceutical companies, provide access to affordable, essential drugs in developing countries.
- In cooperation with the private sector, make available the benefits of new technologies, especially information and communications technologies.

---

ing that current and projected consumption patterns of rich people, coupled with projected demographic changes, lead to resource depletion and undermine the capacity of ecosystems to contribute to human well-being. (See Box 1.1.) In particular, the Millennium Development Goals established by the United Nations in 2000 identify key goals to be achieved on the path to sustainable development. (See Box 1.2.) Achieving most of these—eradicating poverty and hunger, reducing child mortality, improving maternal health, combating HIV/AIDS, eradicating malaria and other diseases, and ensuring environmental sustainability—will require major investments in ecosystem services.

Many private-sector interests also depend on improved ecosystem management. Industries directly dependent on biological resources, such as timber, fishing, or agriculture, have an ever-growing incentive for more effective and efficient management of ecosystem services as demand grows and new sources of supply become increasingly scarce. Far more significant, the condition of ecosystems has become a concern even to companies not directly harvesting biological resources, such as the insurance industry in relation to events associated with climate change. Increased regulation and citizen scrutiny, along with new market incentives and paradigms of corporate stewardship, now drive industries to devote considerable attention to minimizing ecosystem degradation and to factor the condition of the environment into their business strategy. The MA seeks to support and accelerate this process.

Overview of Conceptual Framework

While it is obvious that humans depend on Earth's ecosystems, it is another matter altogether to identify, assess, and undertake practical actions that can enhance well-being without undermining ecosystems. Humans influence, and are influenced by, ecosystems through multiple interacting pathways. Long-term provision of food in a particular region, for example, depends on the characteristics of the local ecosystem and local agricultural practices as well as global climate change, availability of crop genetic resources, access to markets, local income, rate of local population growth, and so forth. Changes at a local scale that may have positive impacts on the local supply of ecosystem services, such as clearing a forest to increase food production, may at the same time have highly detrimental impacts over larger scales: significant loss of forest cover in upstream areas may reduce dry-season water availability downstream, for instance.

Given these complex links between ecosystems and human well-being, a prerequisite for both analysis and action is agreement on a basic conceptual framework. A well-designed framework for either assessment or action provides a logical structure for evaluating the system, ensures that the essential components of the system are addressed as well as the relationships among those components, gives appropriate weight to the different components of the system, and highlights important assumptions and gaps in understanding.

In the case of an ecosystem assessment, an appropriate conceptual framework must cut across spatial dimensions from local to global and across temporal dimensions from the recent past to projections into the next

BOX 1.3   Overarching Questions Guiding the Millennium Ecosystem
Assessment Design

The Millennium Ecosystem Assessment is designed to provide decision-makers with
information to manage ecosystems in a more sustainable manner that will maintain
both biodiversity and the ecosystem services that are essential to human well-being.
Five overarching questions, along with the detailed lists of user needs provided by
convention secretariats and the private sector, guide the issues being assessed:

1. What are the current conditions and trends of ecosystems and their associated
   human well-being?

   - What ecosystems make what contributions to human well-being?
   - How have ecosystems changed in the past and how has this increased or
     reduced their capacity to contribute to human well-being?
     - What thresholds, regime shifts, or irreversible changes have been
       observed?
     - What were the most critical factors affecting the observed changes?
     - What are the costs, benefits, and risks of the observed changes in
       ecosystems, and how have these affected different sectors of society and
       different regions?

2. What are the plausible future changes in ecosystems and in the supply of and
   demand for ecosystem services and the consequent changes in health,
   livelihood, security, and other constituents of well-being?

   - Under what circumstances are thresholds encountered or are regime shifts
     or irreversible changes likely to occur?
   - What are the most critical drivers and factors affecting future changes?
   - What are the costs, benefits, and risks of plausible future human-induced
     changes in ecosystems, and how will these affect different sectors of society
     and different regions?

3. What can we do to enhance well-being and conserve ecosystems? What are
   the strengths and weaknesses of response options, actions, and processes that
   can be considered to realize or avoid specific futures?

   - What are the trade-off implications of the response options?
   - How does inertia in the social and natural systems affect management
     decisions?

4. What are the most robust findings and key uncertainties that affect provision
   of ecosystem services (including the consequent changes in health, livelihood,
   and security) and other management decisions and policy formulations?

5. What tools and methodologies developed and used in the Millennium
   Ecosystem Assessment can strengthen capacity to assess ecosystems, the
   services they provide, their impacts on human well-being, and the implications of response options?

century. It must encompass the accessibility and sustainability of natural resources and systems and their products for the benefit of human societies as well as for the maintenance of these systems in their own right. It must examine how the capacities of ecosystems are being compromised or enhanced, and what mechanisms can be brought to bear to improve the access and delivery of services for human well-being. It must examine all resources simultaneously and in an integrated manner, and must evaluate past and potential future trade-offs and their consequences. To meet all these requirements in a single operational framework for an assessment is a bold venture. Without such comprehensiveness, however, an assessment cannot achieve its goal of understanding the multiple and complex natural and social drivers that are affecting ecosystems and how society can respond in positive ways to maintain ecosystem services that are central to human well-being.

This report describes the conceptual framework that has been developed for the Millennium Ecosystem Assessment. We believe that this framework will be of value to a wide range of analysts and decision-makers who are confronting the challenge of factoring considerations of ecosystems and their services into planning and management, whether it be the design of a business strategy for an agribusiness or the drafting of a national development plan.

The conceptual framework elaborated here has been designed to address a set of core questions developed through extensive interaction with users of the MA, including international conventions, national governments, the private sector, and civil society. (See Box 1.3.)

The basic framework for the MA is shown in Box 1.4. The figure lists the issues that will be addressed in the Millennium Ecosystem Assessment and illustrates their interrelationships. It cannot, of course, portray the complexity of these interactions in their temporal and spatial domains. In particular, the apparent linearity of the relationships between elements of the figure does not fully capture the complex interactions that can occur among them. Given these caveats, the figure and the issues it includes capture the essence of the approach of the MA and provide a framework for structuring the work that needs to be accomplished in the process. Human well-being and poverty reduction are indicated in the upper left-hand box of the conceptual framework diagram. They are placed in this central location to emphasize the primary focus of these issues to the Millennium Ecosystem Assessment.

The MA conceptual framework is designed to assess the consequences of changes in ecosystems for human well-being. It assumes that the cen-

BOX 1.4 Millennium Ecosystem Assessment Conceptual Framework

Changes in factors that indirectly affect ecosystems, such as population, technology, and lifestyle (upper right corner of figure), can lead to changes in factors directly affecting ecosystems, such as the catch of fisheries or the application of fertilizers to increase food production (lower right corner). The resulting changes in the ecosystem (lower left corner) cause the ecosystem services to change and thereby affect human well-being. These interactions can take place at more than one scale and can cross scales. For example, a global market may lead to regional loss of forest cover, which increases flood magnitude along a local stretch of a river. Similarly, the interactions can take place across different time scales. Actions can be taken either to respond to negative changes or to enhance positive changes at almost all points in this framework (black cross bars).

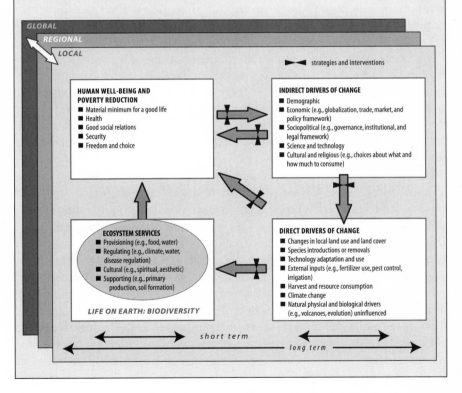

tral components of human well-being—including health, the material minimum for a good life, freedom and choice, health, good social relations, and security—can be linked to the status of the environment. The framework allows examination of the degree to which this is true and under what circumstances. We propose in the work of the MA that the processes maintaining human well-being be the center and keystone of most of the work that is done. In doing this work there is a clear appreciation of the intrinsic value of ecosystems, independent of the services that they provide.

In order to partition the work of the Millennium Ecosystem Assessment, we examine the various services that ecosystems provide and how those services influence human well-being, as well as the forces that have the capacity to alter these services. More specifically, we consider ecosystem services to be the benefits people obtain from ecosystems. For our analysis, we divide these into provisioning, regulating, cultural, and supporting services. These categories overlap extensively, and the purpose is not to establish a taxonomy but rather to ensure that the analysis addresses the entire range of services. There are other ways of categorizing ecosystem services, but the particular approach of the MA seeks to distinguish supporting ecosystem services, which are important for maintaining ecosystems, from those that provide direct benefits to people. Chapter 2 provides a detailed treatment of the role of ecosystems and their services within the MA framework.

Changes in ecosystems affect life on Earth independent of human uses of their services, but we focus particular attention on the consequences of changes in ecosystem services for human well-being. Just as it is not enough to examine a single ecosystem service in isolation from its interaction with other services, so too it is insufficient to focus on only a single attribute of human well-being. Changes in ecosystem services affect many aspects of human well-being. We emphasize in particular the equity dimensions of these changes. Because poor people are often most directly dependent on harvesting ecosystem services, they are often most vulnerable to changes in ecosystems. This framework emphasizes that it is not just the average impact on human well-being that is of interest, but rather the consequences of ecosystem change for different groups of people. We describe the framework used to examine the consequences for human well-being in Chapter 3.

Understanding the factors that are causing ecosystem services to change is essential to designing interventions that can have positive benefits for ecosystems and their services. For convenience of analysis, we consider factors that affect ecosystems directly either through natural processes (such as volcanic eruptions or changes in the sun's energy) or through human actions, such as:

- changes in local land use and land cover;
- modification of river flow;

- species introductions and removals;
- external inputs (such as fertilizer use, pest control, irrigation water);
- discharge of pollutants; or
- harvest of crops, wildlife, or fish.

These factors have had, and are continuing to have, dramatic impacts on ecosystem structure and processes and hence on the services they provide. Many of these factors are in turn driven by demographic, economic, technological, sociopolitical, cultural and religious, physical, biological, and chemical forces that we call indirect drivers of change.

For any given decision-maker, some of these drivers are exogenous, meaning that the individual's decisions will not affect them, while others are endogenous, meaning that decisions directly affect the driver. Thus the small farmer in Africa can decide how much fertilizer to use but cannot influence the global maize price. In contrast, decisions of a finance minister of a major country could influence global maize prices. The role of the direct and indirect drivers of change and their links to decision-makers are examined in Chapter 4.

By depicting a closed loop between its major boxes, the figure in Box 1.4 reflects the existence of feedbacks within the system. In the course of time, indirect drivers are changed not only by long-term general trends, but even more by people's and society's strategies to cope with changing ecosystems in order to maintain well-being. The arrows among the principal contextual boxes of the figure indicate the causal interactions among the components of the system and the general directions of the interactions. The arrows present simplified "if-then" relationships among components: for example, if there is a change in a direct driver, then by definition there will be a change in the ecosystem. In reality, of course, the interactions and their directions are much more complex than depicted.

An important feature of the MA conceptual framework is its multiscale structure, which is depicted in the conceptual framework by the three geographic scales (local, regional, global) and two time scales (short term, long term). The multiscale approach is described in Chapter 5. Briefly, a multiscale assessment contains interlinked assessments conducted at many different geographic scales, which could range from local communities to the entire planet. (See Box 1.5.) It also addresses different time scales, from months or years to decades or centuries. The multiscale component of the MA includes a set of sub-global assessments being conducted within the MA framework, which are now under way or being developed in the

BOX 1.5   Millennium Ecosystem Assessment Sub-global Components

Millennium Ecosystem Assessment (MA) sub-global assessments use the MA conceptual framework; examine conditions, scenarios, and response options; and agree to a set of criteria concerning peer-review, data handling, stakeholder involvement, and intellectual property rights. Each sub-global assessment has significant flexibility in order to meet the needs of its stakeholders effectively. At the same time, the set of sub-global assessments interact extensively to take full advantage of the opportunities for cross-scale integration. The MA includes a set of fully approved and funded sub-global assessments, candidate assessments that have agreed to meet the MA criteria and are now in the design and fundraising stage, and initial ideas for assessments. (See map.) In addition to those shown on the map, close links have been established with ongoing assessments being undertaken by the European Environment Agency and other institutions. More sub-global assessments are expected to join the MA in 2003.

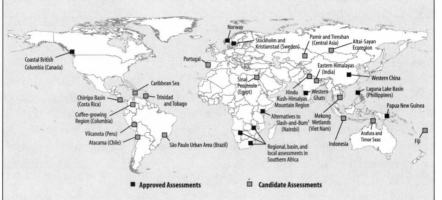

[1] ASB Sites: Sites of the Alternatives to Slash-and-Burn Consortium coordinated by the World Agroforestry Centre (ICRAF). These sites are located in humid tropical forests around the world: Western Brazilian/Peruvian Amazon; Southern Cameroon; Sumatra, Indonesia; Northern Thailand and the Philippines.

Arafura and Timor Seas, Brazil, Canada, the Caribbean Sea, the mountains of Central Asia, Chile, China, Colombia, Costa Rica, Egypt, Fiji, the Hindu Kush-Himalayas, India, Indonesia, Papua New Guinea, Peru, the Philippines, Portugal, Russia, Southern Africa (including Botswana, Mozambique, South Africa, Zambia, and Zimbabwe), Sweden, Trinidad and Tobago, and Viet Nam. In addition, a pilot assessment has been completed in Norway. We expect that other similar sub-global assessments will be established in the next several years.

The choices that people make concerning ecosystems are shaped by what they value in the system. Valuation of ecosystems and their services is unusually difficult, partly because of the intrinsic values that some people ascribe to ecosystems and partly because of the challenge of measuring economic values associated with nonmarketed ecosystem services (Wall et al. 1999; Daily et al. 2000). Typically, economists rely on market prices to provide a measure of the worth of various commodities, but for many ecosystem services, markets simply do not exist. In some cases this is because the costs of transaction and monitoring are too high.

Economic activities affected by ecological interactions involving long geographical distances provide one example of valuation problems. Another example is interactions separated by large temporal distances (the effect of carbon emissions on climate in the distant future, in a world where forward markets do not exist because future generations cannot negotiate with people today). Then there are situations (the atmosphere, aquifers, the open seas) in which the distribution of a resource makes private property rights impossible and so keeps markets from existing. In other cases, ill-specified or unprotected property rights prevent markets from being formed (as happens frequently with mangroves and coral reefs) or make them function incorrectly even if they do get formed. In each of these cases, markets are not providing the correct signals with regard to the value of an ecosystem service. Sound management thus requires alternative means for measuring value as well as policies that can internalize the externalities associated with ecosystem services. Chapter 6 summarizes the various frameworks for thinking about the value of ecosystems and describes how this will be approached in the MA.

Chapter 7 explains the basic analytical approach that can be used in an integrated ecosystem assessment, focusing on the three basic elements of the MA: assessment of current conditions and historical trends; assessment of the consequences of plausible future changes in driving forces; and assessment of the strengths and weaknesses of various response options.

Ultimately, the most important components of the conceptual framework are the black cross bars in the figure in Box 1.4 indicating intervention points where the dynamics of the system can be altered. A major goal of an integrated ecosystem assessment is to provide decision-makers with the information they need to make wise choices concerning these strategies and interventions. This decision-making process is described in Chapter 8.

Much of the work of the Millennium Ecosystem Assessment will involve evaluating interventions that have been successful in the past, as well as proposing novel possibilities that fit the current situation. The MA itself will not recommend specific policies or interventions, since the choice of policies and interventions must be influenced by more than just science. Following the experience of previous assessments, such as the Intergovernmental Panel on Climate Change (IPCC) and the Ozone Assessment, the MA will appraise the strengths and weaknesses of various options, with examples of where and why they have worked. The purpose of a scientific assessment is not to assume a decision-making role by actually selecting the most appropriate option, but rather to contribute to the decision-makers' understanding of the scientific underpinning and implications of various decisions.

The conceptual framework used in the MA differs from the standard environmental impact assessment (EIA) framework in that it places ecosystems and the environment in a central role in the effort to reach development goals. The MA framework is designed to allow the examination of how changes to ecosystems influence human outcomes. The EIA approach, in contrast, focuses on the impacts of human actions on the environment and is designed to explore the relative costs and benefits of various project options. Ecosystems and the environment are treated as externalities in an EIA (affected by development activities), whereas they are internal in the MA framework—something that can be managed sustainably in order to contribute to human development.

The framework also differs from the commonly used pressure-state-impact-response (PSIR) framework by virtue of the feedbacks that it incorporates. The PSIR framework is designed to focus on the impacts of pressures (driving forces) on the environment and the responses that can be taken to alter negative impacts. The MA framework extends the PSIR framework by incorporating the consequences of the environmental impacts on human well-being and as a result turns the relatively linear PSIR framework into a more dynamic system in which environmental changes (the I) can change the human condition and thereby change the pressures (the P).

Equally significant, the MA framework differs from frameworks such as the PSIR or EIA by explicitly including multiscale considerations, as described in the next section. Assessments conducted at different geographic and temporal scales will inevitably focus on different issues and reach different conclusions. No assessment can meet all needs at all scales, but a multiscale framework helps to provide decision-makers with a more complete view of both the issues that need to be addressed and the relative merits of interventions that can be made at different levels of governance.

Each of the four MA Working Groups organizes its work within this conceptual framework. The Condition and Trends Working Group will examine each box of the figure in Box 1.4 (drivers, services, well-being) and their interactions over the past 50 years. The Scenarios Working Group will examine each box and their interactions for different plausible future changes in driving forces, extending out 50 years (and for some variables, 100 years). The Responses Working Group will examine the strategy and intervention points in the figure, which depict options that are available to achieve particular outcomes in the delivery of services from ecosystems. Finally, the Sub-global Working Group will examine all these features (condition, scenarios, and responses) for each of the MA sub-global assessments but at the scale of local communities, river basins, or nations.

The Multiscale Approach

The MA is structured as a multiscale assessment in order to enable its findings to be of greater use at the many levels of decision-making. A global assessment cannot meet the needs of local farmers, nor can a local assessment meet the collective needs of parties to a global convention. A multiscale assessment can also help remedy the biases that are inevitably introduced when an evaluation is done at a single geographic scale. For example, while a national ecosystem assessment might identify substantial national benefits from a particular policy change, a local assessment would be more likely to identify whether that particular community would be a winner or loser as a result of the policy change.

Through the use of a multiscale approach, the findings of the assessment at any scale can in principle be enhanced by virtue of the information and perspectives from other scales. Several factors act together to strengthen the findings of a multiscale assessment. First, a multiscale structure helps to ensure that perspectives or concerns at any given scale are reflected in the analysis and conclusions at other scales. For example, a local community may have quite a different perception of the costs and benefits of various features of the ecosystem than the "global" community. Neither perspective is right or wrong, but a single-scale assessment could miss important differences that could affect the usefulness of various approaches to managing ecosystem change.

Second, a multiscale assessment enables the evaluation of cross-scale factors. Ecosystems are highly differentiated in space and time, and sound management requires careful local planning and action. At the same time, local assessments are insufficient, because some processes are global and because local goods, services, matter, and energy are often transferred across regions. A local assessment of a downstream farming community, for example, would be incomplete without information on upstream activities influencing the community's supply of fresh water.

Finally, a multiscale assessment allows evaluation of the scale-dependence of various actions and policies. Often the aggregate beneficial impacts of a policy change at a national scale may obscure the winners and losers at a local scale. Although differential impacts of change will always exist, the net benefits of actions and policies can be enhanced through more careful assessment of these scale-dependent impacts.

The multiscale framework of the MA is unique among international assessments. Various other global programs include strong regional analyses (such as the *Third Assessment Report* of the IPCC) or produce global findings by aggregating multiple regional assessments (for example, the Global International Waters Assessment, and the *Global Environment Outlook*). The sub-global components of the MA, however, are not just regional analyses or case studies; they are formal assessments undertaken at the sub-global scale, with their own stakeholders, authorizing environments, and user-driven processes.

Types of Knowledge Assessed

Scientific assessments, particularly global assessments, have generally been based on a particular western epistemology (way of knowing), one that often excludes local knowledge, ignores cultural values, and disregards the needs of local communities. Sources such as lay knowledge or practitioners' knowledge tend to be excluded, since assessment procedures often define the information base for an assessment to be the published scientific literature.

Scientists and policy-makers alike have become aware of the need to establish new assessment processes that can accommodate and value these different ways of knowing. For example, a rich body of knowledge concerning the history of ecosystem change and appropriate responses exists within local and traditional knowledge systems, as recognized in principle in the Convention on Biological Diversity. It makes little sense to exclude such knowledge just because it has not been published. Moreover, incorporation of traditional and local knowledge can greatly strengthen the legitimacy of an assessment process in the eyes of indigenous and local communities.

Similarly, substantial knowledge concerning both ecosystems and policy interventions is held within the private sector among the "practitioners" of ecosystem management, yet only a small fraction of this information is ever published in the scientific literature.

Effective incorporation of different types of knowledge in an assessment can both improve the findings and help to increase their adoption by stakeholders if they believe that their information has contributed to those findings. At the same time, no matter what sources of knowledge are incorporated into an assessment, effective mechanisms must be established to judge whether the information provides a sound basis for decisions.

Relatively little experience can be drawn on today of assessment mechanisms that effectively bridge epistemologies. Within the MA, a concerted effort is being made to enable the incorporation of traditional and local knowledge through the establishment of mechanisms for verification even where that knowledge is not first published in peer-reviewed literature. (See Chapter 7.) The MA's multiscale structure provides an unparalleled opportunity to incorporate both traditional and scientific knowledge in the process, since assessments conducted at the smaller scales of individual communities or watersheds will tend to involve much more input of lay and traditional knowledge.

Minimizing Structural Biases

A scientific assessment is a social process to bring the findings of science to bear on the needs of decision-makers. The success of such assessments rests on their saliency, credibility, and legitimacy (Clark and Dickson 1999). Scientific information is salient if it is perceived to be relevant or of value to particular groups who might use it to change management approaches, behavior, or policy decisions. It is credible if peers within the scientific community perceive the scientific and technical information and conclusions to be authoritative and believable. It is legitimate if the process of assembling the information is perceived to be fair and open to input from key political constituencies, such as the private sector, governments, and civil society. The MA has been designed to meet these three basic criteria.

But even the most credible and scientifically unbiased assessment will necessarily give power to some stakeholders at the expense of others. The usefulness of an assessment to different stakeholders is strongly influenced, to begin with, by which stakeholders are involved in choosing its focus. For example, in the face of food insecurity in a particular region, some people may frame the issue as a problem of production and request an assessment of new agricultural technologies for the region, while others may see it as a problem of resource ownership or purchasing power and request an assessment of experience with land redistribution or employment-generating opportunities. Similarly, a global assessment of ecosystem services would naturally examine the role of ecosystems as a source of carbon sequestration, but farmers would be unlikely to select this as an important service unless a mechanism were in place for them to be paid for that sequestration.

The MA, by virtue of its multiscale, multistakeholder structure, will be more neutral with regards to these concerns of focus than other global assessments, but it is not devoid of structural biases. Because its primary authorizing environment is governmental, it will be devoting particular attention to decision-making needs of governments as articulated through the parties to the Convention on Biological Diversity, the Convention to Combat Desertification, the Ramsar Convention on Wetlands, and the Convention on Migratory Species. In addition, although the MA is a multiscale assessment, it will include only about 15 sub-global components. Clearly, an international assessment with thousands of local community components would more strongly reflect the agenda of local communities than an assessment with 15. Thus although the MA for the first time provides a way to increase the input of local or national stakeholders into questions being addressed by an international assessment, it falls short of being scale-neutral and will inevitably focus particular attention on global concerns and questions.

An assessment's usefulness to different stakeholders will also depend on the composition of the scientific community that conducts it. The most effective global assessments, such as the IPCC and the Ozone Assessment, emphasize regional balance of the scientists involved and the involvement of both natural and social sciences. Both regional and disciplinary balance is essential to ensure the credibility and legitimacy of the process. Yet considerable knowledge of ecosystems and their influence on human well-being is held not just in the formal scientific literature but in traditional and local knowledge systems as well. As noted earlier, therefore, the MA is seeking to expand the community of experts conducting the assessment to include local and traditional knowledge. Inevitably, however, while the MA will make an evolutionary step toward more holistic treatment of different ways of knowing the world, the process will still give greater emphasis to peer-reviewed scientific literature.

No assessment can hope to be all things to all people, nor should it be—as it would become highly diffuse. But recognition of the structural biases that exist in any assessment can aid in the interpretation of the findings. And by identifying and describing structural biases, it may be possible to address some of them during the course of the assessment.

Use in Decision-making

Decision-makers confront the full complexity of social-environmental systems with nearly every decision that they take. Scientific assessments, in contrast, have traditionally focused on narrow slices of that complexity. But they increasingly can provide insights into the more complex realities that are at the core of the most difficult choices confronting policy-makers. These tough choices involve trade-offs among different sectors, goals, or time frames. They often involve trade-offs between national and local benefits. And they involve actions to address the structural causes of problems like poverty, not just the symptoms.

Can integrated ecosystem assessments and the information that they make available actually contribute to the real world of decision-making? Despite the growing pressures on ecosystems today, this period in history offers an unprecedented opportunity to modify the development paths being pursued around the world to ones that will secure and sustain human well-being. The last decade has seen progress in understanding how to address environmental and development issues and how to decrease the impact of industry on the environment, but more progress needs to be made in addressing environment and development simultaneously. Today, the world is on the threshold of an era in which integrated environmental management can become a central tool in achieving sustainable development goals. The factors that may have set the stage for this transition include:

- *Advances in science*. Considerable scientific progress has been made in the past several decades in understanding the complex interactions both within ecosystems and between ecosystems, human activities, and human well-being.

- *Advances in information technologies and improved access to information*. Computers and data systems now allow analysts and decision-makers to better monitor ecosystems and predict the consequences of various changes; at the same time, they help provide stakeholders with access to information they need for both decision-making and accountability.

- *Changing paradigms of well-being and poverty*. Historically, human well-being was largely defined in terms of income and consumption; it is now recognized to include the material minimum for a good life, freedom and choice, health, good social relations, security, and peace of mind and spiritual experience.

- *Policy and institutional reform*. Pressures on ecosystems may be exacerbated by misguided policies and institutional arrangements, such as inappropriate subsidies and inequitable patterns of ownership and access to resources. Decision-makes are increasingly aware of the long-term costs of these policies, and many countries are beginning to take steps to reform them.

- *Changing governance*. The relative power of nation-states has diminished with the growth of power and influence of a far more complex array of institutions, including regional governments, multinational companies, the United Nations, and civil society organizations. Many small stakeholders are also increasingly involved in decision-making.

These economic, scientific, institutional, and technological changes have created a new environment for decision-making and action. It is an environment in which multiple users in governments, the private sector, and civil society all have needs for better scientific information and understanding such as that provided through assessments like the MA. And it is an environment in which the general public can make use of information to hold decision-makers accountable. It is also an environment in which it is possible to envision the emergence of new institutional and policy arrangements and changes in rights and access to resources that may be necessary to address fully the challenges of sustainable ecosystem management. In the words of United Nations Secretary-General Kofi Annan in his Millennium Report to the United Nations General Assembly:

> It is impossible to devise effective environmental policy unless it is based on sound scientific information. While major advances in data collection have been made in many areas, large gaps in our knowledge remain. In particular, there has never been a comprehensive global assessment of the world's major ecosystems. The planned Millennium Ecosystem Assessment, a major international collaborative effort to map the health of our planet, is a response to this need (Annan 2000).

Ecosystems and Human Well-being: A Framework for Assessment describes the framework within which such an assessment of the health of the planet can be made. In 2005, the MA will release a series of global assessments undertaken through the application of that framework.

2 Ecosystems and Their Services

Executive Summary

- An ecosystem is a dynamic complex of plant, animal, and microorganism communities and the nonliving environment, interacting as a functional unit. Humans are an integral part of ecosystems.

- A well-defined ecosystem has strong interactions among its components and weak interactions across its boundaries. A useful ecosystem boundary is the place where a number of discontinuities coincide, for instance in the distribution of organisms, soil types, drainage basins, or depth in a water body. At a larger scale, regional and even globally distributed ecosystems can be evaluated based on a commonality of basic structural units.

- Ecosystem services are the benefits people obtain from ecosystems. These include provisioning services such as food and water; regulating services such as flood and disease control; cultural services such as spiritual, recreational, and cultural benefits; and supporting services, such as nutrient cycling, that maintain the conditions for life on Earth.

- Biodiversity is the variability among living organisms. It includes diversity within and among species and diversity within and among ecosystems. Biodiversity is the source of many ecosystem goods, such as food and genetic resources, and changes in biodiversity can influence the supply of ecosystem services.

- People seek many services from ecosystems and thus perceive the condition of an ecosystem in relation to its ability to provide desired services. The ability of ecosystems to deliver services can be assessed by a variety of qualitative and quantitative methods.

- An assessment of the condition of ecosystems, the provision of services, and their relation to human well-being requires an integrated approach. This enables a decision process to determine which service or set of services is valued most highly and how to develop approaches to maintain services by managing the system sustainably.

Introduction

Millions of species populate Earth. The vast majority gain energy to support their metabolism either directly from the sun, in the case of plants, or, in the case of animals and microbes, from other organisms through feeding on plants, predation, parasitism, or decomposition. In the pursuit of life and through their capacity to reproduce, organisms use energy, water, and nutrients. Terrestrial plants obtain water principally from soil, while animals get it mainly from free-standing water in the environment or from their food. Plants obtain most of their nutrients from the soil or water, while animals tend to derive their nutrients from other organisms. Microorganisms are the most versatile, obtaining nutrients from soil, water, their food, or other organisms. Organisms interact with one another in many ways, including competitive, predatory, parasitic, and facilitative ways, such as pollination, seed dispersal, and the provision of habitat.

These fundamental linkages among organisms and their physical and biological environment constitute an interacting and ever-changing system that is known as an ecosystem. Humans are a component of these ecosystems. Indeed, in many regions they are the dominant organism. Whether dominant or not, however, humans depend on ecosystem properties and on the network of interactions among organisms and within and among ecosystems for sustenance, just like all other species.

As organisms interact with each other and their physical environment, they produce, acquire, or decompose biomass and the carbon-based or organic compounds associated with it. They also move minerals from the water, sediment, and soil into and among organisms, and back again into the physical environment. Terrestrial plants also transport water from the soil into the atmosphere. In performing these functions, they provide materials to humans in the form of food, fiber, and building materials and they contribute to the regulation of soil, air, and water quality.

These relationships sound simple in general outline, but they are in fact enormously complex, since each species has unique requirements for life and each species interacts with both the physical and the biological environment. Recent perturbations, driven principally by human activities, have added even greater complexity by changing, to a large degree, the nature of those environments.

Ecosystem Boundaries and Categories

Although the notion of an ecosystem is ancient, ecosystems first became a unit of study less than a century ago, when Arthur Tansley provided an initial scientific conceptualization in 1935 (Tansley 1935) and Raymond Lindeman did the first quantitative study in an ecosystem context in the early 1940s (Lindeman 1942). The first textbook built on the ecosystem concept, written by Eugene Odum, was published in 1953 (Odum 1953). Thus the ecosystem concept, so central to understanding the nature of life on Earth, is actually a relatively new research and management approach.

Tansley's formulation of an ecosystem included "not only the organism-complex, but also the whole complex of physical factors forming what we call the environment" (Tansley 1935:299). He noted that ecosystems "are of the most varied kinds and sizes." The main identifying feature of an ecosystem is that it is indeed a system; its location or size is important, but secondary.

Following Tansley and subsequent developments, we chose to use the definition of an ecosystem adopted by the Convention on Biological Diversity (CBD): "a dynamic complex of plant, animal and micro-organism communities and their nonliving environment interacting as a functional unit" (United Nations 1992:Article 2).

Biodiversity and ecosystems are closely related concepts. Biodiversity is defined by the CBD as "the variability among living organisms from all sources including, *inter alia*, terrestrial, marine and other aquatic ecosystems and the ecological complexes of which they are part; this includes diversity within species, between species and of ecosystems" (United Nations 1992:Article 2). Diversity thus is a structural feature of ecosystems, and the variability among ecosystems is an element of biodiversity. The parties to the convention have endorsed the "ecosystem approach" as their primary framework for action. (See Box 2.1.)

For analysis and assessment, it is important to adopt a pragmatic view of ecosystem boundaries, depending on the questions being asked. In one sense, the entire biosphere of Earth is an ecosystem since the elements interact. At a smaller scale, the guiding principle is that a well-defined ecosystem has strong interactions among its components and weak interactions across its boundaries. (See also Chapter 5.) A practical approach to the spatial delimitation of an ecosystem is to build up a series of overlays of significant factors, mapping the location of discontinuities, such as in the distribution of organisms, the biophysical environment (soil types, drainage basins, depth in a water body), and spatial interactions (home ranges, migration patterns, fluxes of matter). A useful ecosystem boundary is the place where a number of these relative discontinuities coincide. At a larger scale, regional and even globally distributed ecosystems can be evaluated based on the commonality of basic structural units. We use such a framework in the MA for the global analysis of ecosystem properties and changes.

The global assessment being undertaken by the MA is based on 10 categories: marine, coastal, inland water, forest, dryland, island, mountain, polar, cultivated, and urban. (See Box 2.2.) These categories are not ecosystems themselves, but each contains a number of ecosystems. The MA reporting categories are not mutually exclusive: their boundaries can

BOX 2.1   The Ecosystem Approach: A Bridge Between the Environment and
          Human Well-being

The concept of an ecosystem provides a valuable framework for analyzing and acting on the linkages between people and their environment. For that reason, the
ecosystem approach has been endorsed by the Convention on Biological Diversity
(CBD) and the Millennium Ecosystem Assessment (MA) conceptual framework is
entirely consistent with this approach. The CBD defines the ecosystem approach as
follows:

> The Ecosystem Approach is a strategy for the integrated management of
> land, water and living resources that promotes conservation and sustain-
> able use in an equitable way. Thus, the application of the ecosystem
> approach will help to reach a balance of the three objectives of the Con-
> vention: conservation; sustainable use; and the fair and equitable sharing
> of the benefits arising out of the utilization of genetic resources. An eco-
> system approach is based on the application of appropriate scientific meth-
> odologies focused on levels of biological organization, which encompass
> the essential structure, processes, functions and interactions among or-
> ganisms and their environment. It recognizes that humans, with their
> cultural diversity, are an integral component of many ecosystems.

According to the CBD, the term ecosystem can refer to any functioning unit at
any scale. This approach requires adaptive management to deal with the complex
and dynamic nature of ecosystems and the absence of complete knowledge or understanding of their functioning. It does not preclude other management and conservation approaches, such as biosphere reserves, protected areas, and single-species
conservation programs, or other approaches carried out under existing national policy
and legislative frameworks; rather, it could integrate all these approaches and other
methodologies to deal with complex situations. As described in the CBD, there is
no single way to implement the ecosystem approach, as it depends on local, provincial, national, regional, and global conditions.

The conceptual framework of the MA provides a useful assessment structure that
can contribute to the implementation of the CBD's ecosystem approach. By way of
analogy, decision-makers would not make a decision about financial policy in a
country without examining the condition of the economic system, since information on the economy of a single sector such as manufacturing would be insufficient.
The same applies to ecological systems or ecosystems. Decisions can be improved by
considering the interactions among the parts of the system. For instance, the drain-
ing of wetlands may increase food production, but sound decisions also require information on whether the potential added costs associated with the increased risk
of downstream flooding or other changes in ecosystem services might outweigh those
benefits.

and do overlap. Ecosystems within each category share a suite of biological, climatic, and social factors that tend to differ across categories. More specifically, there generally is greater similarity within than between each category in:

- climatic conditions;
- geophysical conditions;
- dominant use by humans;
- surface cover (based on type of vegetative cover in terrestrial ecosystems or on fresh water, brackish water, or salt water in aquatic ecosystems);
- species composition; and
- resource management systems and institutions.

The factors characterizing ecosystems in each category are highly interrelated. Thus, for example, grasslands are found in many areas where potential evaporation exceeds precipitation. Grasslands, in turn, tend to be used by humans either as rangeland or for agricultural purposes. The areas used for rangeland tend to have pastoral, sometimes nomadic, resource management systems. Thus these factors—high potential evaporation relative to precipitation, grassland cover, use for livestock, and pastoral or nomadic management systems—tend to be found together. (This is typical of the dryland system category in Box 2.2.)

We use overlapping categories in the global MA analysis because this better reflects real-world biological, geophysical, social, and economic interactions, particularly at these relatively large scales. For example, an important issue for ecosystems and human well-being in forested regions relates to the impact of forest harvest or conversion on the timing, quantity, and quality of water runoff. Given the importance of this interaction, it is helpful to analyze an area dominated by forest land cover as a single ecosystem even if it contains some freshwater and agricultural areas within it, rather than analyzing the forest, agriculture, and freshwater ecosystems separately, since this allows for a more holistic analysis of these interactions.

Ecosystem Services

Ecosystem services are the benefits people obtain from ecosystems. This definition is derived from two other commonly referenced and representative definitions:

> Ecosystem services are the conditions and processes through which natural ecosystems, and the species that make them up, sustain and fulfill human

BOX 2.2   Reporting Categories Used in the Millennium Ecosystem Assessment

Social and ecological systems can be categorized in an infinite number of ways. For the purposes of reporting the global Millennium Ecosystem Assessment (MA) findings, we have developed a practical, tractable, sufficiently rich classification based on 10 systems. Thus, for example, the MA will report on "forest systems," defined to be areas with at least 40 percent canopy (tree) cover. Using this approach, a forest system will contain a variety of different types of ecosystems, such as freshwater ecosystems, agroecosystems, and so forth. But all areas within the boundaries of the forest system as defined here will tend to share a suite of biological, climatic, and social factors, so the system categories provide a useful framework for analyzing the consequences of ecosystem change for human well-being. Because the boundaries of these reporting categories overlap, any place on Earth may fall into more than one category. Thus a wetland ecosystem in a coastal region, for instance, may be examined both in the MA analysis of "coastal systems" as well as in the analysis of "inland water systems."

The following table lists the basic boundary definitions that will be used in the global MA analysis. In a number of cases the MA will also examine conditions and changes in ecosystems with reference to more than one boundary definition. For example, although we use a boundary of 40 percent tree (canopy) cover as our basic definition of the forest category, another widely accepted definition of "forests" is at least 10 percent canopy cover.

Millennium Ecosystem Assessment Reporting Categories

Category	Central Concept	Boundary Limits for Mapping
Marine	Ocean, with fishing typically a major driver of change	Marine areas where the sea is deeper than 50 meters.
Coastal	Interface between ocean and land, extending seawards to about the middle of the continental shelf and inland to include all areas strongly influenced by the proximity to the ocean	Area between 50 meters below mean sea level and 50 meters above the high tide level or extending landward to a distance 100 kilometers from shore. Includes coral reefs, intertidal zones, estuaries, coastal aquaculture, and seagrass communities.
Inland water	Permanent water bodies inland from the coastal zone, and areas whose ecology and use are dominated by the permanent, seasonal, or intermittent occurrence of flooded conditions	Rivers, lakes, floodplains, reservoirs, and wetlands; includes inland saline systems. Note that the Ramsar Convention considers "wetlands" to include both inland water and coastal categories.
Forest	Lands dominated by trees; often used for timber, fuelwood, and non-timber forest products	A canopy cover of at least 40 percent by woody plants taller than 5 meters. The existence of many other definitions is acknowledged, and other limits (such as crown cover greater than 10 percent, as used by the Food and Agriculture Organization of the United Nations) will also be reported. Includes temporarily cut-over forests and plantations; excludes orchards and agroforests where the main products are food crops.

BOX 2.2   continued

Millennium Ecosystem Assessment Reporting Categories

Category	Central Concept	Boundary Limits for Mapping
Dryland	Lands where plant production is limited by water availability; the dominant uses are large mammal herbivory, including livestock grazing, and cultivation	Drylands as defined by the Convention to Combat Desertification, namely lands where annual precipitation is less than two thirds of potential evaporation, from dry subhumid areas (ratio ranges 0.50–0.65), through semiarid, arid, and hyper-arid (ratio <0.05), but excluding polar areas; drylands include cultivated lands, scrublands, shrublands, grasslands, semi-deserts, and true deserts.
Island	Lands isolated by surrounding water, with a high proportion of coast to hinterland	As defined by the Alliance of Small Island States
Mountain	Steep and high lands	As defined by Mountain Watch using criteria based on elevation alone, and at lower elevation, on a combination of elevation, slope, and local elevation range. Specifically, elevation >2,500 meters, elevation 1,500–2,500 meters and slope >2 degrees, elevation 1,000–1,500 meters and slope >5 degrees or local elevation range (7 kilometers radius) >300 meters, elevation 300–1,000 meters and local elevation range (7 kilometers radius) >300 meters, isolated inner basins and plateaus less than 25 square kilometers extent that are surrounded by mountains.
Polar	High-latitude systems frozen for most of the year	Includes ice caps, areas underlain by permafrost, tundra, polar deserts, and polar coastal areas. Excludes high-altitude cold systems in low latitudes.
Cultivated	Lands dominated by domesticated plant species, used for and substantially changed by crop, agroforestry, or aquaculture production	Areas in which at least 30 percent of the landscape comes under cultivation in any particular year. Includes orchards, agroforestry, and integrated agriculture-aquaculture systems.
Urban	Built environments with a high human density	Known human settlements with a population of 5,000 or more, with boundaries delineated by observing persistent night-time lights or by inferring areal extent in the cases where such observations are absent.

life. They maintain biodiversity and the production of ecosystem goods, such as seafood, forage timber, biomass fuels, natural fiber, and many pharmaceuticals, industrial products, and their precursors (Daily 1997b:3).

Ecosystem goods (such as food) and services (such as waste assimilation) represent the benefits human populations derive, directly or indirectly, from ecosystem functions (Costanza et al. 1997:253).

The MA definition follows Costanza and his colleagues in including both natural and human-modified ecosystems as sources of ecosystem services, and it follows Daily in using the term "services" to encompass both the tangible and the intangible benefits humans obtain from ecosystems, which are sometimes separated into "goods" and "services" respectively.

Like the term ecosystem itself, the concept of ecosystem services is relatively recent—it was first used in the late 1960s (e.g., King 1966; Helliwell 1969). Research on ecosystem services has grown dramatically within the last decade (e.g., Costanza et al. 1997; Daily 1997a; Daily et al. 2000; de Groot et al. 2002).

It is common practice in economics both to refer to goods and services separately and to include the two concepts under the term services. Although "goods," "services," and "cultural services" are often treated separately for ease of understanding, for the MA we consider all these benefits together as "ecosystem services" because it is sometimes difficult to determine whether a benefit provided by an ecosystem is a "good" or a "service." Also, when people refer to "ecosystem goods and services," cultural values and other intangible benefits are sometimes forgotten.

Ecosystem services have been categorized in a number of different ways, including by:

- functional groupings, such as regulation, carrier, habitat, production, and information services (Lobo 2001; de Groot et al. 2002);

- organizational groupings, such as services that are associated with certain species, that regulate some exogenous input, or that are related to the organization of biotic entities (Norberg 1999); and

- descriptive groupings, such as renewable resource goods, nonrenewable resource goods, physical structure services, biotic services, biogeochemical services, information services, and social and cultural services (Moberg and Folke 1999).

For operational purposes, we will classify ecosystem services along functional lines within the MA, using categories of provisioning, regulating, cultural, and supporting services. (See Figure 2.1.) We recognize that some of the categories overlap.

Provisioning Services

These are the products obtained from ecosystems, including:

- *Food and fiber.* This includes the vast range of food products derived from plants, animals, and microbes, as well as materials such as wood, jute, hemp, silk, and many other products derived from ecosystems.

- *Fuel.* Wood, dung, and other biological materials serve as sources of energy.

FIGURE 2.1   Ecosystem Services

Ecosystem services are the benefits people obtain from ecosystems. These include provisioning, regulating, and cultural services that directly affect people and supporting services needed to maintain the other services.

Provisioning Services	Regulating Services	Cultural Services
Products obtained from ecosystems	*Benefits obtained from regulation of ecosystem processes*	*Nonmaterial benefits obtained from ecosystems*
■ Food	■ Climate regulation	■ Spiritual and religious
■ Fresh water	■ Disease regulation	■ Recreation and ecotourism
■ Fuelwood	■ Water regulation	■ Aesthetic
■ Fiber	■ Water purification	■ Inspirational
■ Biochemicals	■ Pollination	■ Educational
■ Genetic resources		■ Sense of place
		■ Cultural heritage

Supporting Services
Services necessary for the production of all other ecosystem services
■ Soil formation   ■ Nutrient cycling   ■ Primary production

- *Genetic resources*. This includes the genes and genetic information used for animal and plant breeding and biotechnology.

- *Biochemicals, natural medicines, and pharmaceuticals*. Many medicines, biocides, food additives such as alginates, and biological materials are derived from ecosystems.

- *Ornamental resources*. Animal products, such as skins and shells, and flowers are used as ornaments, although the value of these resources is often culturally determined. This is an example of linkages between the categories of ecosystem services.

- *Fresh water*. Fresh water is another example of linkages between categories—in this case, between provisioning and regulating services.

Regulating Services

These are the benefits obtained from the regulation of ecosystem processes, including:

- *Air quality maintenance*. Ecosystems both contribute chemicals to and extract chemicals from the atmosphere, influencing many aspects of air quality.

- *Climate regulation.* Ecosystems influence climate both locally and globally. For example, at a local scale, changes in land cover can affect both temperature and precipitation. At the global scale, ecosystems play an important role in climate by either sequestering or emitting greenhouse gases.

- *Water regulation.* The timing and magnitude of runoff, flooding, and aquifer recharge can be strongly influenced by changes in land cover, including, in particular, alterations that change the water storage potential of the system, such as the conversion of wetlands or the replacement of forests with croplands or croplands with urban areas.

- *Erosion control.* Vegetative cover plays an important role in soil retention and the prevention of landslides.

- *Water purification and waste treatment.* Ecosystems can be a source of impurities in fresh water but also can help to filter out and decompose organic wastes introduced into inland waters and coastal and marine ecosystems.

- *Regulation of human diseases.* Changes in ecosystems can directly change the abundance of human pathogens, such as cholera, and can alter the abundance of disease vectors, such as mosquitoes.

- *Biological control.* Ecosystem changes affect the prevalence of crop and livestock pests and diseases.

- *Pollination.* Ecosystem changes affect the distribution, abundance, and effectiveness of pollinators.

- *Storm protection.* The presence of coastal ecosystems such as mangroves and coral reefs can dramatically reduce the damage caused by hurricanes or large waves.

Cultural Services

These are the nonmaterial benefits people obtain from ecosystems through spiritual enrichment, cognitive development, reflection, recreation, and aesthetic experiences, including:

- *Cultural diversity.* The diversity of ecosystems is one factor influencing the diversity of cultures.

- *Spiritual and religious values.* Many religions attach spiritual and religious values to ecosystems or their components.

- *Knowledge systems* (traditional and formal). Ecosystems influence the types of knowledge systems developed by different cultures.

- *Educational values*. Ecosystems and their components and processes provide the basis for both formal and informal education in many societies.
- *Inspiration*. Ecosystems provide a rich source of inspiration for art, folklore, national symbols, architecture, and advertising.
- *Aesthetic values*. Many people find beauty or aesthetic value in various aspects of ecosystems, as reflected in the support for parks, "scenic drives," and the selection of housing locations.
- *Social relations*. Ecosystems influence the types of social relations that are established in particular cultures. Fishing societies, for example, differ in many respects in their social relations from nomadic herding or agricultural societies.
- *Sense of place*. Many people value the "sense of place" that is associated with recognized features of their environment, including aspects of the ecosystem.
- *Cultural heritage values*. Many societies place high value on the maintenance of either historically important landscapes ("cultural landscapes") or culturally significant species.
- *Recreation and ecotourism*. People often choose where to spend their leisure time based in part on the characteristics of the natural or cultivated landscapes in a particular area.

Cultural services are tightly bound to human values and behavior, as well as to human institutions and patterns of social, economic, and political organization. Thus perceptions of cultural services are more likely to differ among individuals and communities than, say, perceptions of the importance of food production. The issue of valuing ecosystem services is addressed in Chapter 6.

Supporting Services

Supporting services are those that are necessary for the production of all other ecosystem services. They differ from provisioning, regulating, and cultural services in that their impacts on people are either indirect or occur over a very long time, whereas changes in the other categories have relatively direct and short-term impacts on people. (Some services, like erosion control, can be categorized as both a supporting and a regulating service, depending on the time scale and immediacy of their impact on people.) For example, humans do not directly use soil formation services, although changes in this would indirectly affect people through the impact on the provisioning service of food production. Similarly, climate regulation is categorized as a regulating service since ecosystem changes can have an impact on local or global climate over time scales relevant to human decision-making (decades or centuries), whereas the production of oxygen gas (through photosynthesis) is categorized as a supporting service since any impacts on the concentration of oxygen in the atmosphere would only occur over an extremely long time. Some other examples of supporting services are primary production, production of atmospheric oxygen, soil formation and retention, nutrient cycling, water cycling, and provisioning of habitat.

A Multisectoral Approach

Every part of Earth produces a bundle of ecosystem services. (See Box 2.3.) Human interventions can increase some services, though often at the expense of other ones. Thus human interventions have dramatically increased food provisioning services through the spread of agricultural technologies, although this has resulted in changes to other services such as water regulation. For this reason, a multisectoral approach is essential to fully evaluate changes in ecosystem services and their impacts on people. The multisectoral approach examines the supply and condition of each ecosystem service as well as the interactions among them. The MA has adopted just such an approach.

When assessing ecosystem services, it is often convenient to bound the analysis spatially and temporally with reference to the ecosystem service or services being examined. Thus a river basin is often the most valuable ecosystem scale for examining changes in water services, while a particular agroecological zone may be more appropriate for assessing changes in crop production. When looking at interactions among services, the combination of services provided by an ecosystem, or the variety of services drawn on by a society, the question of boundaries becomes more complex. Issues of boundaries, scale, and habitat heterogeneity are important and are dealt with in greater detail in Chapter 5.

Biodiversity and Ecosystem Services

Habitat modification, invasion, and many other factors are leading to changes in biodiversity across many taxa within most ecosystems. Recently, theoretical and empirical work has identified linkages between changes in biodiversity and the way ecosystems function (Schulze and Mooney 1993; Loreau et al. 2002). The MA will address how ecosystem services are affected by such linkages.

BOX 2.3   Analysis of Ecosystem Services

Any region of Earth produces a set of services that in turn influences human well-being. It also receives flows of energy, water, organisms, pollutants, and other materials from adjacent regions and releases similar materials into those regions. Various strategies and interventions influence the quantity and quality of the services provided.

An ecosystem is typically composed of a number of different regions, such as forest, agriculture, and urban areas, each of which produces a different bundle of services. In an ecosystem assessment, both the production of services from each area and the flows of materials between areas must be assessed.

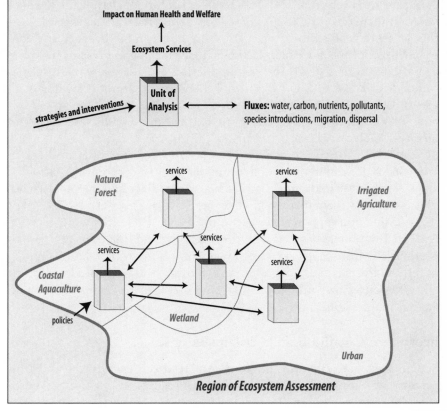

Among the most important factors identified is the degree of functional redundancy found within an ecosystem. This indicates the substitutability of species within functional groups in an ecosystem such that the impact created by the loss of one or more species is compensated for by others (Naeem 1998). For example, in many ecosystems there are several species that fix nitrogen (known as a functional group of species). If the loss of any one of them is compensated for by the growth of others and there is no overall loss in nitrogen fixation, then there is functional redundancy in that ecosystem.

Some species make unique or singular contributions to ecosystem functioning, however, and therefore their loss is of greater concern (Walker 1992). Small changes in the biodiversity of diverse systems may lead to only small changes in the functioning of an ecosystem, including its production of services, providing no species with unique roles are lost (Jones et al. 1994; Power et al. 1996). But the possibility of significant losses of function increases as more species are lost and as redundancy is reduced— that is, there is an asymptotic relationship between biodiversity and ecosystem functioning. For example, the high diversity of South African fynbos ecosystems ensures steady rates of production because many plant species can compensate for losses by growing when others cannot (Cowling et al. 1994). Greater redundancy represents greater insurance that an ecosystem will continue to provide both higher and more predictable levels of services (Yachi and Loreau 1999).

The MA will seek to evaluate biodiversity and potential declines in biodiversity for different ecosystems under a set of different scenarios for plausible changes in driving forces. This work will extend previous studies that developed scenarios for biodiversity change (Sala et al. 2000). For provisioning and supporting services, the MA will identify which ecosystem functions are associated with these services and link their response to declining biodiversity, using the fundamental asymptotic relationship between biodiversity and ecosystem functioning. Both magnitudes and stability responses to biodiversity loss can be considered using this fundamental relationship.

Ecosystem Condition and Sustainable Use

People seek multiple and different services from ecosystems and thus perceive the condition of an ecosystem in relation to its ability to provide the services desired. The ability of ecosystems to deliver particular services can be assessed separately with various methods and measures. An adequate assessment of the condition of ecosystems, the provision of services, and their implications for human well-being (see Chapter 3) requires an integrated approach. With such an assessment in hand, a decision process (see Chapter 8) can then determine which set of services are valued most highly (see Chapter 6) and can manage the system in a sustainable way.

In a narrow sense, the sustainability of the production of a particular ecosystem service can refer simply to whether the biological potential of the ecosystem to sustain the yield of that service (such as food production) is being maintained. Thus a fish provision service is sustainable if the surplus but not the resource base is harvested, and if the fish's habitat is not degraded by human activities. In the MA, we use the term "sustained yield management" to refer to the management and yield of an individual resource or ecosystem service.

More generally, however, sustainability is used in the context of "sustainable development" to refer to a pattern of development that meets current needs without diminishing prospects for future generations. We use sustainability, and sustainable management, to refer to this goal of ensuring that a wide range of services from a particular ecosystem is sustained.

The MA will consider criteria and methods to provide an integrated approach to ecosystem assessment. The condition and sustainability of each category of ecosystem services is evaluated in somewhat different ways, although in general a full assessment of any service requires considerations of stocks, flows, and resilience.

Condition of Provisioning Services

The flows of provisioning services do not accurately reflect their condition, since a given flow may or may not be sustainable over the long term. The flow is typically measured in terms of biophysical production, such as kilograms of maize per hectare or tons of tuna landings. The provisioning of ecological goods such as food, fuelwood, or fiber, depends both on the flow and the "stock" of the good, just as is the case with manufactured goods. (In economics, "stock" refers to the total merchandise kept on hand by a merchant; in this section, we use "stock" in its economic sense to show how considerations of ecosystem goods can be incorporated into the economic framework of stocks and flows.) The quantity of goods sold by a manufacturer (the flow), for example, is an incomplete measure of a factory's productivity, since it could come from either the production of new goods or the depletion of built-up stocks. Indeed, production of biological resources has often been maintained in the short term at a higher rate than its sustainable yield. In the long term, the production of overharvested resources will fall.

Marine fisheries provide examples of an ecosystem service being degraded even while output has been temporarily maintained or increased by more intensive harvesting. Numerous fisheries around the world have been overharvested, exhibiting a general pattern of rapid growth in landings (production) followed by the eventual collapse of the fishery. (See Box 2.4.) Similar patterns can be found with virtually all other provisioning services.

Agricultural production, for example, can be maintained through the addition of fertilizers and through new crop varieties even while the productive potential of the ecosystem is degraded through soil erosion. Some 40 percent of agricultural land has been strongly or very strongly degraded in the past 50 years by erosion, salinization, compaction, nutrient depletion, biological degradation, or pollution even while overall global food production has increased (WRI et al. 2000). So long as manufactured capital can compensate for losses of the natural capital of the ecosystem, agricultural production can be maintained. In this case, however, manufactured and natural capital are not perfectly substitutable, and once a critical level of soil degradation is reached, agricultural output will decline. A complete accounting of the condition of food production would reveal that it has been degraded because the underlying capability of the ecosystem to maintain production has been degraded.

Historically, it has not been common for environmental or resource assessments to include measures of the productive potential of biological resources when monitoring the condition of the resource. Thus although all countries have considerable information on the production of grain, fisheries, and timber, relatively little is known about the actual condition of these services since the productive potential of the resource has rarely been evaluated. The Pilot Analysis of Global Ecosystems, which was prepared by the World Resources Institute and the International Food Policy Research Institute to assist in the MA design, attempted to provide a more complete assessment of the condition of ecosystem services along these lines (Matthews et al. 2000; Revenga et al. 2000; White et al. 2000; Wood et al. 2000).

Condition of Regulating, Cultural, and Supporting Services

In the case of regulating services, as opposed to provisioning services, the level of "production" is generally not relevant. Instead the condition of the service depends more on whether the ecosystem's capability to regulate a particular service has been enhanced or diminished. Thus if forest clearance in a region has resulted in decreased precipitation and this has had harmful consequences for people, the condition of that regulatory service has been degraded.

BOX 2.4   Collapse of the Atlantic Cod Fishery

The Atlantic cod stocks off the east coast of Newfoundland collapsed in 1992, forcing the closure of the fishery after hundreds of years of exploitation. Until the late 1950s, the fishery was exploited by migratory seasonal fleets and resident inshore small-scale fishers. From the late 1950s, offshore bottom trawlers began exploiting the deeper part of the stock, leading to a large catch increase and a strong decline in the underlying biomass. Internationally agreed quotas in the early 1970s and, following the declaration by Canada of an Exclusive Fishing Zone in 1977, national quota systems ultimately failed to arrest and reverse the decline.

Two factors that contributed to the collapse of the cod stock were the shift to heavy fishing offshore and the use of fishery assessment methods that relied too much on scientific sampling and models based on the relatively limited time series and geographical coverage of the offshore part of the fish stocks. Traditional inshore fishers, whose landings account for one third to one half of the total, had noticed the decline in landings even before the mid-1980s, ahead of the scientists involved in fisheries assessment work but these observations could not be used in stock assessments because of technical difficulties in converting the catches into a suitable form. Finlayson (1994) noted that "science will confer the status of 'valid' only on very specific forms of data presented in a very specialized format."

Northern Cod Off Newfoundland, Canada (NAFO *area* 2J3KL)

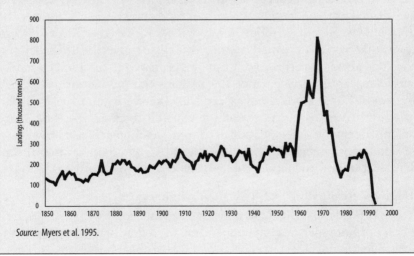

Source: Myers et al. 1995.

The evaluation of the condition of cultural services is more difficult. Some cultural services are linked to a provisioning service (such as recreational fishing or hunting) that can serve as a proxy measure of the cultural service. But in most cases no such proxy exists. Moreover, unlike provisioning or regulating services, assessing the condition of cultural services depends heavily on either direct or indirect human use of the service. For example, the condition of a regulating service such as water quality might be high even if humans are not using the clean water produced, but an ecosystem provides cultural services only if there are people who value the cultural heritage associated with it.

Information about the condition of cultural services can be obtained by identifying the specific features of the ecosystem that are of cultural, spiritual, or aesthetic significance and then examining trends in those features. For example, salmon are a totemic or revered species in almost all parts of the world where they are found, and thus the degradation of wild salmon stocks represents degradation of a cultural service provided by the ecosystem. But cultural service information such as this would be difficult to obtain and to quantify: tigers, for instance, remain totemic species even in areas where they have been extinct for decades. Recognizing that the concept of cultural services is relatively new, the MA will explore methods for evaluating the condition and value of these services.

Supporting services maintain the conditions for life on Earth but may affect people only indirectly (by supporting the production of another service, as soil formation supports food production) or over very long time periods (such as the role of ecosystems in producing oxygen). Because the link to human benefits is indirect, as opposed to the other ecosystem services just discussed, a normative scale for assessing the condition of a service is not always practical. For example, primary production is a fundamental supporting service, since life requires the production of organic compounds. But if global primary production were to increase by 5 percent over the next century, it would be difficult to categorize the change as an enhancement or degradation of the service, though it certainly would be a significant change. In such cases the MA will report on the current biophysical state (production, flux, and stocks) of the supporting service.

Variability, Resilience, and Thresholds in Services

Whenever possible, individuals and governments generally invest in various types of insurance that can buffer human welfare against natural variability. Such investments may be as basic as establishing limited stores of food, medicine, and potable water for disaster relief to more elaborate investments such as building dams, levies, and canals to guard against 100-year floods. How, when, and where to invest in such insurance requires assessing not just mean levels of stocks and flows of ecosystem services but also their dynamics or, more specifically, their variability and stability.

Three characteristics of ecosystem services are important in such an assessment: ecosystem variability, resilience, and thresholds. (See Box 2.5.) There are many other properties of stability in dynamic systems (such as resistance, sensitivity, persistence, reliability, predictability, and so forth), but the MA will limit its focus to these three important and well-studied stability properties.

Variability in ecosystem services consists of changes in stocks or flows over time due to stochastic, intrinsic, and extrinsic factors, all of which must be disentangled to understand system behavior properly. Stochastic variability is due to random or uncontrolled factors creating variability that is often considered background or "white" noise in system behavior.

BOX 2.5   Dynamics and Stability in Ecosystem Services

This figure illustrates the level of provisioning of an ecosystem service that has been perturbed twice. Hypothetically, such a service exhibits stochastic (random or uncontrolled) and inherent variability (fluctuations above and below the two horizontal lines, which represent different system states). The system recovers after the first perturbation, with its resilience being measured by the duration of the recovery phase or return time to its first state. Note that crossing the threshold of the second state does not cause a shift when in the first state. The second perturbation causes the service to cross the second threshold, which leads to a regime shift or catastrophic change to an alternative stable state. The long dashed lines illustrate two thresholds. Only when a system crosses a threshold does it switch to an alternate state.

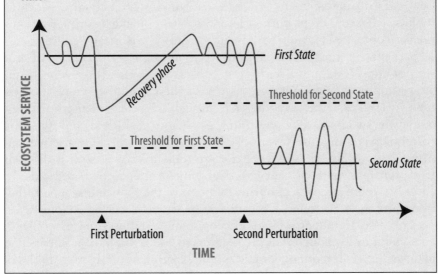

In contrast, intrinsic (inherent) variability is due to the structural properties of an ecosystem, such as oscillations in systems where predation or disease regulate the number of animals. Examples of extrinsic variability, due to forces outside the system, include seasonality in temperate systems and longer-term climate systems such as El Niño–La Niña cycles.

Resilience is most often considered a measure of the ability of a system to return to its original state after a perturbation—a deviation in conditions that is outside the range experienced over a decade or more, such as a large-scale fire or an unusually severe drought. When the duration of the recovery phase is short in comparison to other systems, the system is considered to be more resilient than the others.

Thresholds or breakpoints in ecosystems represent dramatic, usually sudden (less than a decade) deviations from average system behavior. Such dramatic shifts—also known as regime shifts, catastrophic change, or entering alternative stable states—are often primed by a steady change in internal or external conditions that increases a system's susceptibility to being triggered to enter an alternative state (Scheffer et al. 2001; Carpenter 2003). For example, on a global scale, small, steady increases in global warming may lead to a sudden reorganization of Earth's ocean circulation patterns (Broecker 1997). On a local scale, the increase in grazing animals by ranchers or herders may be responsible for shifts in steppe (grass-dominated) to tundra (moss-dominated) ecosystems (Zimov et al. 1995).

While management goals are often conceived in terms of stocks and flows, reducing system variability and improving predictability are often key parts of management strategies. Examples of such interventions include irrigating crops during droughts, using biocides during pest outbreaks, controlled burning to prevent catastrophic fires, and culling herds to prevent a population explosion. Maintaining forests to prevent erosion or coral reefs to prevent wave impacts in the face of severe storms are examples of managing ecosystems for their insurance value. Ecosystem variability is often addressed through a variety of methods, but management aimed at maintaining ecosystem resilience and avoiding thresholds is sometimes overlooked. In part this is because the mechanisms responsible for such behavior are seldom known, so it is difficult to design management that can deal with resilience or thresholds. In addition, there are no accurate assessments of the probability of perturbations, and the time frame over which such events occur is too long.

The costs to human welfare of ecosystem deviation from its norms of behavior, however, are often severe, thus its inclusion in assessments and management is important. The MA will examine not only magnitudes of ecosystem stocks and flows as they are related to ecosystem goods and services, but also their stability properties. Much of this will be done by extrapolation from expert assessment of paleo records (for instance, climate records derived from ice cores) and historical records (such as long-term fisheries, forestry, or agricultural records) to obtain guidelines on the norms of system variability, resilience, known thresholds, and the environmental stresses that cause ecosystems to be triggered by perturbations to enter into alternative states.

Ecosystem Health and Other Related Concepts

Ecosystem health is a concept that has often been applied to the evaluation of ecosystems (Rapport et al. 1995). This has become a subdiscipline in the life sciences, with its own journals and professional organizations, such as the International Society for Ecosystem Health (ISEH) and the Aquatic Ecosystem Health and Management Society. The term is used sometimes to mean the links between ecosystems and human health. For example, the mission of ISEH is to "encourage the understanding of the critical linkages between human activity, ecological change and human health" (Rapport et al. 1999:83). It is also used to refer to the health of the ecosystem itself: "an ecological system is healthy...if it is stable and sustainable—that is, if it is active and maintains its organization and autonomy over time and is resilient to stress" (Costanza et al. 1992:9).

This concept has generated debate and alternative approaches within the scientific literature (e.g., Reid 1996; de Leo and Levin 1997). One method measures health as a departure from some preferred (often "natural") state. Another, which is consistent with the approach used in the MA to examine the condition of ecosystem services, relates health to the ability of an ecosystem within its surrounding landscape to continue to provide a particular set of services. This considers whether the ecosystem and its external inputs (such as energy or fertilizer) are sustainable in the long term as well as whether the ecosystem can withstand or recover from perturbations (resistance and resilience, respectively) and similar issues.

The concept of ecosystem health is important both within the research community and as a means of communicating information about ecosystems to the general public. Although the MA has not adopted ecosystem health as its primary organizational framework, the concept could be usefully applied within an assessment that used the MA framework.

Several other concepts will also inform the MA without being adopted as organizational frameworks. For instance, ecosystem integrity has been defined as "the maintenance of the community structure and function characteristic of a particular locale or deemed satisfactory to society" (Cairns 1977:56) or "the capability of supporting and maintaining a balanced, integrated, adaptive community of organisms having species composition, diversity, and functional organization comparable to that of natural habitats of the region" (Karr and Dudley 1981:171). Another example is the "ecological footprint," which expresses the impact of human activity on ecosystems in terms of areas required to provide the services used by an individual or community.

Substitution of Services

Substitutes are available for some ecosystem services, although often the cost of a technological substitution will be high and it may not replace all the services lost. For example, water treatment plants can now substitute for ecosystems in providing clean drinking water, although this may be expensive and will not overcome the impacts of water pollution on other components of the ecosystem and the services they provide. Another outcome of substitution is that often the individuals gaining the benefits are not those who originally benefited from the ecosystem services. For example, local coastal fish production can be replaced by shrimp aquaculture in tropical regions, but the individuals making a living from capture fisheries are not those who would profit from the new shrimp aquaculture facilities.

Therefore, a full assessment of ecosystems and their services must consider:

- information on the cost of a substitute,
- the opportunity cost of maintaining the service,
- cross-service costs and impacts, and
- the distributional impacts of any substitution.

3 Ecosystems and Human Well-being

Executive Summary

- Human well-being has several key components: the basic material needs for a good life, freedom and choice, health, good social relations, and personal security. Well-being exists on a continuum with poverty, which has been defined as "pronounced deprivation in well-being."

- How well-being and ill-being, or poverty, are expressed and experienced is context- and situation-dependent, reflecting local social and personal factors such as geography, ecology, age, gender, and culture. These concepts are complex and value-laden.

- Ecosystems are essential for human well-being through their provisioning, regulating, cultural, and supporting services. Evidence in recent decades of escalating human impacts on ecological systems worldwide raises concerns about the consequences of ecosystem changes for human well-being.

- Human well-being can be enhanced through sustainable human interaction with ecosystems with the support of appropriate instruments, institutions, organizations, and technology. Creation of these through participation and transparency may contribute to people's freedoms and choices and to increased economic, social, and ecological security.

- Some believe that the problems from the depletion and degradation of ecological capital can be largely overcome by the substitution of physical and human capital. Others believe that there are more significant limits to such substitutions. The scope for substitutions varies by socioeconomic status.

- We identify direct and indirect pathways between ecosystem change and human well-being, whether it be positive or negative. Indirect effects are characterized by more complex webs of causation, involving social, economic, and political threads. Threshold points exist, beyond which rapid changes to human well-being can occur.

- Indigent, poorly resourced, and otherwise disadvantaged communities are generally the most vulnerable to adverse ecosystem change. Spirals, both positive and negative, can occur for any population, but the poor are more vulnerable.

- Functioning institutions are vital to enable equitable access to ecosystem services. Institutions sometimes fail or remain undeveloped because of powerful individuals or groups. Bodies that mediate the distribution of goods and services may also be appropriated for the benefit of powerful minorities.

- For poor people, the greatest gains in well-being will occur through more equitable and secure access to ecosystem services. In the long run, the rich can contribute greatly to human well-being by reducing their substantial impacts on ecosystems and by facilitating greater access to ecosystem services by the poor.

- We argue ecological security warrants recognition as a sixth freedom of equal weight with participative freedom, economic facilities, social opportunities, transparency guarantees, and protective security.

Introduction

As noted in earlier chapters, the impacts of human activities on ecosystems have increased rapidly in the last few decades. While the majority of these can be considered beneficial to human well-being, there is growing evidence of adverse effects. Clear analysis of these undesirable impacts and their consequences for people has been difficult because of the numerous other causes of ecosystem change that operate and interact at different social, geographical, and temporal scales. For some people, especially those buffered by relative affluence, the problem is scarcely visible—or at least accorded low priority. Yet millions of others experience every day the detrimental consequences of ecosystem changes.

Consideration of purely local and overt environmental deficiencies, such as visible pollution, is no longer a sufficient framework to assess the relationship between the environment and human well-being. The recently evident larger-scale changes to the world's ecosystems must also be looked at closely (McMichael 2001).

The dependence of humans on ecosystem services reflects directly the profound co-evolutionary processes that underlie the origins of Earth's biosphere. The biosphere and its ecosystems provide life support to all species, as described in Chapter 2. Further, the biosphere is itself the product of life on Earth. The composition of the atmosphere and soil, the cycling of nutrients through waterways, and many other ecological assets are all the result of living processes—and all are maintained and replenished by living ecosystems.

The effects of adverse ecosystem changes on human well-being can be classed as direct and indirect. Direct effects occur with some immediacy, through locally identifiable biological or ecological pathways. For example, impairment of the water-cleansing capacity of wetlands may adversely affect those who drink that water. Building dams can increase mosquito-breeding and thus the transmission of malaria. The deforestation of hillsides can expose downstream communities to the hazards of flooding.

Indirect effects take a toll on well-being through more complex webs of causation, including through social, economic, and political routes. Some may take decades to have an impact. For example, where farmlands under irrigation become saline, crop yields are reduced; this in turn may affect human nutritional security, child growth and development, and susceptibility to infectious diseases. Beyond threshold points, limited or degraded supplies of fresh water may exacerbate political tensions, impair local economic activity (and livelihoods)—including industry—and reduce aesthetic amenity. These dynamic, interacting processes jeopardize various aspects of human well-being.

The impacts of adverse ecosystem change do not fall evenly on human populations. Indigent, poorly resourced, and otherwise disadvantaged communities are generally the most vulnerable. Further, many poor rural populations rely disproportionately on the integrity and functions of local ecosystems and are likely to lack the means to import ecosystem services. Impoverishment as a result of adverse ecosystem change may sometimes lead to a downwards spiral for such people. In all instances, the ability to achieve well-being is reduced by the diminished availability of ecosystem services.

Key Components of Human Well-being

There have been many formulations and definitions of human well-being (Alkire 2002). Most commentators would agree that it includes basic material needs for a good life, the experience of freedom, health, personal security, and good social relations. Together, these provide the conditions for physical, social, psychological, and spiritual fulfillment.

A distinction is sometimes made between the determinants of or means to well-being and its constituents—that is, well-being as an end (Dasgupta 2001). In other words, well-being is experiential, what people value being and doing. The determinants are sometimes expressed as commodity inputs, many of which are provided by ecosystem services. They include food, fiber, fuel, clean water, materials for shelter, marketed crops, livestock, forest products, and minerals. Enabling physical, environmental, and social conditions and access—for example, to resources and space—

are also relevant as determinants of or means to well-being. Viewed within this frame, some key elements of well-being can be both determinants and constituents. For example, education and health can be both ends in themselves and the means to experience well-being.

There is widespread agreement that well-being and poverty are the two extremes of a multidimensional continuum. In fact, the *World Development Report 2000/01* defined poverty as "the pronounced deprivation of well-being" (World Bank 2001).

How well-being and ill-being, or poverty, are expressed and experienced is context- and situation-dependent, reflecting local social and personal factors such as geography, ecology, age, gender, and culture (Prescott-Allen 2001). Although these concepts are recognized as complex and value-laden, some elements are nevertheless widespread—if not universal. This was evident in the "voices of the poor" research (Narayan et al. 1999; 2000), in which poor people in 23 countries were asked to reflect, analyze, and express their ideas of the bad and the good life. The respondents stressed many aspects, including the importance of secure and adequate livelihoods, cultural and spiritual activities, and the ability to provide for their children. Repeatedly, they indicated five linked components (see Figure 3.1):

- the necessary material for a good life (including secure and adequate livelihoods, income and assets, enough food at all times, shelter, furniture, clothing, and access to goods);
- health (including being strong, feeling well, and having a healthy physical environment);
- good social relations (including social cohesion, mutual respect, good gender and family relations, and the ability to help others and provide for children);
- security (including secure access to natural and other resources, safety of person and possessions, and living in a predictable and controllable environment with security from natural and human-made disasters); and
- freedom and choice (including having control over what happens and being able to achieve what a person values doing or being).

These five dimensions reinforce each other, whether positively or negatively. A change in one often brings about changes in the others. The shaded space in Figure 3.1 represents the experience of living and being—including stress, pain, and anxiety in the bad life and peace of mind and spiritual experience in the good life.

FIGURE 3.1   The Main Dimensions of Well-being and its Obverse, Ill-being

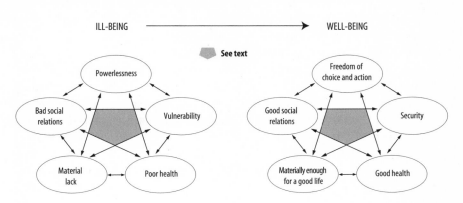

In this multidimensional formulation, there are negative and positive webs of interactions. On the side of ill-being and the bad life, the double-headed arrows represent negative directions of causality: for example, poor people are vulnerable to sickness, which in turn makes them poorer; bad social relations make people vulnerable to shocks, which in turn deepens material poverty and so on; and all of these contribute to powerlessness. On the side of well-being and the good life, having materially enough facilitates physical strength, enabling a better livelihood, while good social relations can provide security against stresses and shocks. In turn, security is likely to increase material well-being and so on. And all of these enhance freedom of choice and action.

Overall, development can thus be seen as the enhancement of well-being. It entails transitions for those who are deprived—from conditions of ill-being or the "bad life" to well-being or the "good life."

One condition for personal well-being is the capability to adapt and achieve that which individuals value doing and being in situations of dynamic change. At the social level this may contribute to conflicts, necessitating trade-offs between the well-being of different individuals and groups. Trade-offs may occur when, for example, material capital is accumulated at a cost of environmental security or cultural or spiritual values. This also has a temporal dimension concerning the well-being of others in the future.

Addressing these issues leads into the sphere of values. This is a realm for decision-makers. The Millennium Ecosystem Assessment (MA) does not take a position, but we note that one proposed approach to these conflicts and trade-offs is a framework that combines concepts of equity, sustainability, livelihood, capability, and ecosystem stewardship. These are related to a value-based notion of well-being in which socially and eco-logically responsible behavior plays a part (Chambers 1997a). This in turn relates to the negative and positive effects of individuals' lives, actions, and non-actions on ecosystems and on other people—both now and in the future. Negative effects manifest especially through the unsustainable consumption of resources, the degradation of ecosystems, and the many impacts of the behaviors of people who are richer and more powerful on those who are poorer and weaker. Positive effects include sustainable rela-tionships between people and ecosystems, as well as the provision and enhancement of present and future livelihoods, capabilities, and human well-being.

Linkages between Ecosystem Services and Human Well-being

These formulations recognize that the relationship of ecosystem condi-tions and the flow of services to the well-being of groups of people as well as individuals is diverse and complex. Further, it changes over time. Many ecosystem changes are planned, but many are inadvertent consequences of other human activities. Human interventions in nature have had unex-pected and surprising consequences, some of which have harmed and fur-ther impoverished those who are disadvantaged. Equitable and sustain-able well-being depends heavily on links with ecosystem services and on who gains and who loses over time from their use. As noted in Chapter 2, the MA has identified four major categories of ecosystem services that bear directly on human well-being: provisioning, regulating, cultural, and supporting services.

The provisioning function of ecosystems supplies goods and other ser-vices that sustain various aspects of human well-being. By the same token, shortages of food, fiber, and other products have adverse effects on human well-being, via both direct and indirect pathways. Adverse impacts on livelihoods are of particular importance. In both social and environmen-tal contexts, livelihood sustainability has three aspects:

- a livelihood is sustainable "when it can cope with and recover from stresses and shocks and maintain or enhance its capabilities and assets both now and in the future" (DFID 1999);
- a livelihood is sustainable in a social context when it enhances or does not diminish the livelihoods of others; and

- a livelihood is sustainable when it does not deplete or disrupt ecosystems to the prejudice of the livelihoods and well-being of others now or in the future.

Biodiversity is fundamental to many ecosystem services. For example, it provides sustainability and resilience vital for the livelihoods and coping strategies of many people, especially the rural poor. They often obtain ecosystem services, and thereby reduce their vulnerability, through diverse and complex mixes of activities over the seasons. For them, biodiversity has a stabilizing and buffering function. It provides multiple sources of ecosystem services, as well as fallback options for food and other resources when times are bad (Davies 1996; Chambers 1997b; Carney 1998; Ellis 1998; Koziell 1998; Scoones 1998; Neffjes 2000).

The regulating functions of ecosystems also affect human well-being in multiple ways. These include the purification of air, fresh water, reduced flooding or drought, stabilization of local and regional climate, and checks and balances that control the range and transmission of certain diseases, including some that are vector-borne. Without these regulatory functions, the varied populations of human and animal life are inconceivable. Thus changes to an ecosystem's regulatory function may have consequences for human health and other components of well-being.

Ecosystems also have many consequences for human well-being through the cultural services they provide—through, for example, totemic species, sacred groves, trees, scenic landscapes, geological formations, or rivers and lakes. These attributes and functions of ecosystems influence the aesthetic, recreational, educational, cultural, and spiritual aspects of human experience. Many changes to these ecosystems, through processes of disruption, contamination, depletion, and extinction, therefore have negative impacts on cultural life and human experience.

Supporting services are essential for sustaining each of the other three ecosystem services. Thus the link between supporting services and human well-being occurs indirectly.

The diverse links between ecosystem services and the determinants and constituents of human well-being are illustrated in Figure 3.2. The spatial and temporal forms of these links, as well as their complexity, vary greatly. Some relationships are immediate; others are lagged. For instance, impairment of food production causes hunger today and malnutrition before long, bringing lassitude, impaired ability to concentrate and learn, and increased vulnerability to infectious diseases. Examples of longer time-lags include the clearing of mangroves, which impairs the replenishment

FIGURE 3.2   Ecosystem Services and Their Links to Human Well-being

Ecosystem services are the benefits people obtain from ecosystems. These include provisioning, regulating, and cultural services, which directly affect people, and supporting services needed to maintain the other services. Changes in these services affect human well-being through impacts on security, the necessary material for a good life, health, and social and cultural relations. These constituents of well-being are in turn influenced by and have an influence on the freedoms and choices available to people. (See also Duraiappah 2002.)

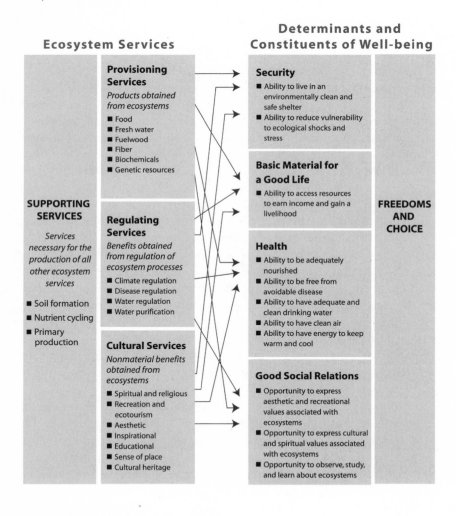

of fish stocks (Naylor et al. 2000), salinization created by badly managed shrimp aquaculture, depletion of groundwater for irrigation, and the impact of introduced species.

Some larger-scale environmental stresses heighten tensions, leading to possible conflict, and threaten well-being by causing health problems (Homer-Dixon 1994). For example, Ethiopia and the Sudan, which are both upstream of Egypt, increasingly need the Nile's water for their own crops. Worldwide, approximately 40 percent of the world's population, living in 80 countries, now faces some level of water shortage (Gleick 2000). The construction of large dams, though of benefit through irrigation and power generation, can create new stresses—particularly in developing countries—by leading to increased levels of schistosomiasis (Fenwick et al. 1981) or displacing people through flooding (Roy 1999; World Commission on Dams 2000).

The dual challenge for society is thus to retain and, indeed, sustain a sufficient level of ecosystem services in a way that contributes to the enhancement of human well-being and the reduction of poverty. Explicit recognition of these links (see Box 3.1) and of substitutability among the various forms of capital will help policy-makers and other stakeholders to make informed decisions. Those, in turn, may produce the most efficient and equitable outcome.

Substitutability and Well-being

Ecosystem services can be conceptualized as flows parallel to those from physical and human capital. Some of these services can also be partially replaced by using physical capital. For instance, limited amounts of clean air and water can be obtained by air-conditioning a space or by using water filters. In other words, partial substitutability exists for at least some ecosystem services. Some commentators believe that the problems from the depletion and degradation of ecological capital can be largely overcome by the accumulation of knowledge and of manufactured and human capital. There are limits to substitution possibilities, however, and the scope for substitutions varies by social, economic, and cultural conditions.

In fact, the substitution possibilities open to a community depend critically on economic status. A resource can be a luxury for others even while it is a necessity for some. Politically, commercial demand can easily outrank local needs, especially under nondemocratic regimes. If local biodiversity is lost, ecotourists can go somewhere else, where it still exists. International public opinion, not to mention pressure from a country's

---

**BOX 3.1   Environment, Population, Poverty, and Well-being:
A Complex Relationship**

The downward spiral that links environment, poverty, health, and well-being is complex. Both poverty and environmental degradation, via independent pathways, jeopardize well-being and health.

Some commentators maintain that an approximate inverted U-shaped relationship exists between income and environmental degradation. That is, as the average income of a population increases, many forms of environmental degradation initially increase before the availability of wealth, literacy, and regulatory institutions combine to reduce the problem (Grossman and Krueger 1995).

The poor, however, derive their sustenance and livelihoods from healthy ecosystems such as grasslands, forests, and cropland. Why do they degrade the very assets that are the source of their own present and future incomes? Does their poverty make them barter the future for the present? Studies in the past decade from many parts of the developing world show that this usually happens when local social institutions that govern the use of "the commons" break down (Chopra et al. 1990; Chopra and Gulati 2001; Jodha 2001; Markandya 2001). This may be due to the operation of a combination of factors, including commercialization, population pressure, and bad governance. When appropriate sets of property rights are put in force, the process can be contained.

Most of the documented examples of an inverted U-shaped relationship refer to local pollution such as river or air pollution. In contrast, the indices for many of today's larger-scale environmental problems (such as greenhouse gas emissions and the release of activated nitrogen) display a continuous increase (Vitousek et al. 1997; Butler 2000). These are problems of the "global commons" (Dasgupta 1996; Buck 1998) for which there is not yet clear feedback in terms of perceived consequences that influence the richer populations. Finding appropriate interventions for them will require the agency of global institutions.

---

elite, is often at best tepid. Local needs are frequently overridden by outsiders' demands (Guard and Masaiganah 1997).

When wetlands, forests, and woodlands are converted (for agriculture, for example, or urban development), local communities may suffer. For them, and especially the poorest, there are few substitutes or choices. For privileged others, whose "ecological footprints" dwarf those of the poor and weak (Wackernagel and Rees 1995), there are often substitutes—something else, often somewhere else. Issues of common and conflicting interests and of reducing demands can be expected to surface. The question that may arise is whether long-term and secure well-being for the world's relatively affluent people will lie in living more lightly on Earth while ensuring a better life and a fairer share of ecosystem services for those who are poor and deprived. In any case, there are vital policy questions about how to achieve well-being for all, and especially for those who experience it least.

Balancing Priorities: Present Versus Future

The relationship between ecosystem change and human well-being has both current and future dimensions. The overexploitation of ecosystems may temporarily increase material well-being and alleviate poverty, yet it may prove unsustainable. That is, to solve today's pressing problems, society is often tempted to deplete tomorrow's ecological resource base. This can jeopardize future well-being and, in some cases, even survival.

The World Commission on Environment and Development first proposed a now widely accepted definition of sustainable development as "development that meets the needs of the present without compromising the ability of future generations to meet their own needs" (WCED 1987:43). That is, each generation should bequeath to its successor at least as large a productive base as it inherited. Thus the concept of sustainable development incorporates not only intragenerational but also intergenerational equity.

In practice, can the present generation be expected to pursue sustainable development policies? After all, parents care about both the current and future well-being of their children. Since their children's well-being will depend upon the well-being of their grandchildren—and that of their grandchildren will in turn depend upon their great-grandchildren's, and so on—parents will tend to take at least some account of the interests of their distant descendants, even if they are directly interested only in their children.

Such individual concerns find a reflection in societal preferences only when prevailing property rights and other institutional structures take them into account. This is rarely the case. Instead, bad or weakly functioning institutions not only permit adverse consequences for human well-being from past and present actions but also hold no one to account. Often, the damage to ecosystems is the result of elite and powerful groups, both domestic and international, extracting short-term values for quick gains, thereby overriding the often longer-term interest of individuals and local communities (Jepson et al. 2001). If property rights to local ecosystems are ill defined or inadequately protected, such actions can have long-term adverse effects on ecosystem services that no one is responsible for.

Just as such actions can adversely affect contemporaries, they can have unintended consequences over time. For example, fish farms created by clearing mangroves can benefit economically the company that has created the farms, but the action will inflict future damage on those who would otherwise have depended on the mangroves for provisioning, regulating, supporting, and cultural services (Gilbert and Janssen 1998; Ong 2002).

Institutions and Freedoms

Earlier sections demonstrated how many of the constituents or determinants of well-being were directly or indirectly provided by ecosystem services. It has also been shown that ecosystem services are not infinite and are subject to scarcity. Although there are potentials for substitutability with other forms of capital, thresholds exist beyond which substitutes are not possible. For example, while many pharmaceuticals can be produced synthetically, the therapeutic potential of extinct, undiscovered species can never be developed.

Scarcity and the chance to add value provide powerful incentives for individuals or groups to try to gain privileged access and rights-of-use to many ecosystems and their services. They do this by influencing the political, economic, and social institutions that govern their access, management, and use (Ostrom 1990; Acheson 1993; Alston et al. 1997; Ensmiger 1997).

Institutions—formal and informal—mediate the link between ecosystem services and the constituents and determinants of human well-being. For example, institutions for community forest management in India have successfully facilitated access to forest products for local communities (Chopra and Dasgupta 2002).

In most cases, inequitable distribution of or access to ecosystems and their services occurs when formal or informal institutions break down (Binswager 1989; Jaganathan 1989; Duraiappah 1998). This happens either when institutions do not exist or when they are inefficient or ineffective. There are many reasons for institutional failure. Commonly, powerful individuals or groups prevent the establishment of institutions. Existing bodies that mediate the distribution of goods and services may also be appropriated for the benefit of powerful minorities. Agricultural subsidies in western industrial countries are an example of this.

Creating, revising, and modifying institutions is a social process. Certain preconditions, or "freedoms," are necessary to ensure that this process is equitable and fair. These freedoms, by permitting a fair and equitable social process, play a critical role in preventing or mitigating institutional failure. Five freedoms that have been identified are participative freedom, economic facilities, social opportunities, transparency guarantees, and protective security (Jordan 1996; Sen 1999; Chopra and Duraiappah in press). For example, access by the poor to credit at reasonable interest rates—the provision of an economic facility—has been facilitated by microcredit schemes, such as the Grameen Bank, a formal institution (Yunus 1998).

We have, in this chapter, added a sixth freedom to the five just listed— ecological security. We define this as the minimum level of ecological stock (an ecological safety net), defined by respective communities through an open and participatory process, that is required to provide the supporting services needed to ensure a sustainable flow of provisioning, regulating, and cultural ecosystem services. We stress that ecosystems and their services are not only instrumental for improving well-being, but are also constitutive elements of well-being. For example, microbiologically adequate water, needed for good health, may also be valued for aspects such as its purity and ease of access.

Contrary to the view that some of these freedoms are luxuries, deferrable until some level of macroeconomic growth has been achieved, we argue that they are complementary, rather than substitutes. Social, political, economic, and ecological freedoms are essential if equity, fairness, justice, and choice are to be addressed. In order to take advantage of economic facilities, for instance, it is essential to have some social opportunities, such as health and education, available (Drèze and Sen 2002). In a similar fashion, it is necessary to have participative freedom and transparency guarantees if ecological security is to truly benefit local communities.

These six freedoms provide the space that allows individuals to define their rights—legal, political, social, ecological—and to create institutions to protect and oversee a fair and equitable distribution of these rights for all members of society. In this manner, individuals, especially the poor, are given the ability to make their own choices for self-determination. This process allows them to become agents of change.

Conclusion

The well-being of present and future human populations depends on ecologically sustainable and socially equitable ways of living in the world. In determining how to achieve these, value judgments have to be made concerning equity and ecosystem stewardship. These are the sphere of policy-makers. Depending on context, decision-makers are faced with questions of who gains and who loses in rights, access, and the ability to enjoy ecosystem services.

Toward these ends, and toward the reduction of poverty, an essential step is fuller understanding of the myriad ways in which human activities and well-being are related to ecosystem changes and services. Such understandings will always be needed to inform and support responsible and farsighted governance. It is implicit in this chapter that enhancing those understandings will be an essential and permanent part of human endeavor. To achieve sustainable well-being for all will be a perennial challenge. And in the constant flux and interaction of ecosystems and people, no answers can ever be final.

4 *Drivers of Change in Ecosystems and Their Services*

EXECUTIVE SUMMARY

- Understanding the factors that cause changes in ecosystems and ecosystem services is essential to the design of interventions that enhance positive and minimize negative impacts.

- A driver is any natural or human-induced factor that directly or indirectly causes a change in an ecosystem. A direct driver unequivocally influences ecosystem processes and can therefore be identified and measured to differing degrees of accuracy. An indirect driver operates more diffusely, often by altering one or more direct drivers, and its influence is established by understanding its effect on direct drivers.

- Decision-makers influence some drivers and are influenced by other drivers. The first are the endogenous drivers and the latter are the exogenous ones. Conceptually, decisions are made at three organizational levels: by individuals and small groups at the local level who directly alter some part of the ecosystem; by public and private decision-makers at municipal, provincial, and national levels; and by public and private decision-makers at the international level. In reality, however, the distinction between these levels is often diffuse and difficult to define.

- The degree to which a driver is outside the influence of a decision-making process depends to some extent on the temporal scale. Some factors may be exogenous in the short run but subject to change by a decision-maker over longer periods.

- Local decision-makers can directly influence the choice of technology, changes in land use, and external inputs but have little control over prices and markets, property rights, technology development, or the local climate. National or regional decision-makers have more control over many indirect drivers, such as macroeconomic policy, technology development, property rights, trade barriers, prices, and markets.

- The indirect drivers of change are primarily demographic, economic, sociopolitical, scientific and technological, and cultural and religious. The interaction of several of these drivers in turn affects the overall level of resource consumption and disparities in consumption within and between countries. Clearly these drivers are changing: population and the global economy are growing, there are major advances in information technology and biotechnology, and the world is becoming more interconnected. Changes in these drivers are projected to increase the demand for food, fiber, clean water, and energy, which will in turn affect the direct drivers. The direct drivers are primarily physical, chemical, and biological, such as land cover change, climate change, air and water pollution, irrigation, use of fertilizers, harvesting, and the introduction of alien invasive species.

- Any decision can have consequences external to the decision framework. These are called externalities because they are not part of the decision-making calculus. Externalities can have positive or negative effects. The effect of an externality is seldom confined to the environs of the decision-maker. External effects extend to other parts of the ecosystem and even to other ecosystems. It is possible for individually unimportant external effects to have dramatic regional and global consequences when many local decision-makers simultaneously take decisions with similar unintended consequences.

- Multiple, interacting drivers cause changes in ecosystem services. There are functional interdependencies between and among the indirect and direct drivers of change, and, in turn, changes in ecological services lead to feedbacks on the drivers of changes in ecological services. Synergetic driver combinations are very common. The many processes of globalization are leading to new forms of interactions among drivers of changes in ecosystem services.

Introduction

A broad range of factors lead directly and indirectly to changes in ecosystems, ecosystem services, and human well-being. Many ecosystem changes are intended or unintended consequences of human decisions and the ensuing actions. The drivers of those changes may be well defined, such as grain prices or local rainfall, but they may also involve more complex and diffuse interactions arising from institutional or cultural influences. Understanding the factors that cause these changes in ecosystems and ecosystem services is essential to designing interventions that enhance positive and minimize negative impacts.

Here, as in many parts of the Millennium Ecosystem Assessment (MA), the first challenge is to find terms that mean the same thing to many different users. The term "driver," for example, is used widely in the ecological and other natural sciences but seldom used by economists. And even when the term is used, different meanings exist. The MA defines

---

BOX 4.1   Typologies of Drivers

Several typologies of drivers were considered for the Millennium Ecosystem Assessment conceptual framework—primary versus proximate, anthropogenic versus biophysical, dependent versus independent, primary versus secondary. The proximate and primary driver terminology, for example, is widely used in the land use change and climate change literature (e.g., Turner II et al. 1995; IPCC 2002). Proximate and primary drivers are conceptually similar to direct and indirect drivers respectively, but tend to be used when analyzing specific spatial processes in which the human intent (primary) is linked with actual physical actions (proximate). The explicit cross-scale linkages and inclusion of physical activities of this typology made it too complex, however, for characterizing the drivers in the Millennium Ecosystem Assessment conceptual framework. Other typologies have been developed for specific purposes and have their limitations. The distinction between direct and indirect drivers, in contrast, provides an opportunity to include highly diverse types of drivers and seemed acceptable to the broadest possible community.

---

driver in the broadest possible sense: any natural or human-induced factor that directly or indirectly causes a change in an ecosystem.

The approach adopted here is to distinguish between direct and indirect drivers. (See Box 4.1.) A direct driver unequivocally influences ecosystem processes and therefore can be identified and measured to differing degrees of accuracy. Indirect drivers operate more diffusely, from a distance, often by altering one or more direct drivers. An indirect driver can seldom be identified through direction observation of the ecosystem; its influence is established by understanding its effect on a direct driver.

A decision-maker can influence certain driving forces (the endogenous drivers) but not others (the exogenous drivers). Endogenous drivers are thus under the direct control of a decision-maker at a certain level, while exogenous drivers are not. The MA explicitly focuses on who controls specific drivers. This helps to explain the role of responses in describing, understanding, and projecting changes in ecosystems, ecosystem services, and human well-being.

Consider, for example, the case of wheat production in Europe. A wheat farmer in southern France can vary the amount of nitrogenous fertilizer to apply but has no influence on the price received for the wheat. Policymakers in the European Union, however, can influence the price of wheat received by that farmer by imposing or eliminating wheat trade restrictions. As the time and space scales expand, more drivers become endogenous; that is, a different set of decision-makers has influence over the drivers. This distinction is especially important in identifying intervention points and strategies.

Another key point is that any decision can have consequences external to the decision framework. These consequences are called externalities because they are not part of the decision-making calculus. Externalities can have positive or negative effects. For example, a decision to plow a dry field for crop production might result in substantial particulate matter blowing into a nearby village, with negative health effects. But it is also possible to have positive externalities. A beekeeper might be motivated by the profits to be made from selling honey, for instance, but neighboring orchards could produce more apples because of enhanced pollination arising from the presence of the bees.

Previous Approaches on the Factors of Change

During the late 1960s and early 1970s, debates began about the factors that lead humans to have adverse effects on the biophysical environment. A number of "root" causes were asserted: religion (White 1967), common property institutions (McCay and Jentoft 1998), and capitalism and colonialism (O'Connor 1988). But none of these hypotheses of dominant cause could sustain empirical scrutiny. The IPAT formulation (Impacts = Population x Affluence x Technology) was an initial attempt to move beyond simple arguments about single causes by acknowledging:

- that there are multiple human drivers of environmental change,
- that their effects are multiplicative rather than additive,
- that increases in one driver can sometimes be mitigated by changes in another driver, and
- that assessing the effects of human drivers requires both theory and empirical evidence.

For a history of IPAT and related arguments about drivers, see Dietz and Rosa (1994).

IPAT continues to be used in discussions of the drivers of environmental change (e.g., Waggoner and Ausubel 2002), and the IPAT accounting framework finds productive use in industrial ecology (Chertow 2001). However, formulations that build on IPAT are emerging. The impact of population growth and affluence on consumption continues to be examined. A variety of studies demonstrate that population size has an effect on impacts but sometimes is less important than other factors (e.g., Palloni

1994; Rudel and Roper 1997; York et al. 2003). A substantial literature examines the effects of affluence on environmental impact (reviewed in Stern 1998; Nordstrom and Vaughan 1999), including a number of analyses that suggest that such effects depend strongly on context (Roberts and Grimes 1997). Research on drivers deploys the full repertoire of available methodologies, including statistical analyses, case studies, and simulation, and the literature is growing in both size and sophistication.

Over the last decade, the approach has been further refined in many assessment models by adding such factors as specific sociopolitical, biophysical, and cultural drivers. But these top-down approaches to environmental change still rely heavily on highly aggregated drivers, the value of which has recently been questioned (e.g., Barbier 2000; Contreras-Hermosilla 2000; Barrett et al. 2001; Indian National Academy of Sciences et al. 2001; Lambin et al. 2001; Myers and Kent 2001; van Beers and de Moor 2001; Young 2002). For example, in a statistical analysis of the causes of deforestation, Geist and Lambin (2002) show that different local and regional drivers play an important role. But perhaps the most important recent advance in understanding is the elucidation of a broader variety of interacting drivers that become more important in the local context.

The individual importance of global drivers cannot be assessed in a simple way. There is no clear hierarchy of drivers that encompasses cause and effect. Individuals and societies try to influence their environment and fulfill their needs by evaluating expected outcomes. If undesired impacts are foreseen, mitigating decisions can be made. This approach is made operational most clearly in the Driver-Pressure-State-Impact-Response (DPSIR) scheme that was developed by the Organisation for Economic Co-operation and Development (OECD InterFutures Study Team 1979).

Many assessments have followed this approach, at least in part. For example, the Intergovernmental Panel for Climate Change structured its assessment along these lines—activities > emissions > concentration > climate change > impacts > mitigation and adaptation responses (IPCC 2002)—recognizing that responses in turn alter activities (mitigation) and impacts (adaptation). The conceptual framework is a closed loop and displays different interactions between drivers and components. In the MA, determining trade-offs and synergies between different decisions and other responses will be central. This requires that the assessment examines carefully the interactions of drivers at specific scale levels and over varying spatial, temporal, and organizational dimensions.

Recent advances in integrated assessment (e.g., Alcamo et al. 1998; Stafford-Smith and Reynolds 2002) and comprehensive analyses of environmental problems (e.g., Petschel-Held et al. 1999; Ostrom et al. 2002) have shown that analyzing causes of environmental change requires a multiscale and multidimensional assessment of major components of the system and their dynamics and interactions. An appreciation of the feedbacks, synergies, and trade-offs among these components in the past improves understanding of current conditions and enhances the ability to project future outcomes and intervention options.

Drivers: An Overview

In the MA, key elements of drivers that are assessed include:

- an explicit recognition of the role of decision-makers at different levels who directly or indirectly affect ecosystems and their services;
- identification of drivers that influence these decision-makers;
- the specific temporal, spatial, and organizational scale dependencies of these drivers; and
- the specific linkages and interactions among drivers.

The MA approach assumes that decisions are made at three organizational levels:

- by individuals and small groups at the local level (such as fields and forest stands) who directly alter some part of the ecosystem;
- by public and private decision-makers at regional levels (the municipal, provincial, and national level); and
- by international conventions and multilateral agreements that operate at the global level.

For global drivers, we recognize that there is no explicit global governing body. The United Nations proceeds, for example, through consensus building between national governments. And in reality, of course, the distinction between these three levels is often diffuse and difficult to assess.

Today a fairly consistent, agreed-upon list of global or "big-picture" drivers that change ecosystems, ecosystem services, and human well-being has emerged. Many of these are used as inputs to models that project future energy and land use (e.g., Nakícenovíc et al. 2000). However, many of these models use global aggregates, and distinct local and regional patterns in these drivers are not captured. The major global driving forces used in many assessments, which the MA uses as a basis for analysis, are:

- demographic drivers;
- economic drivers;
- sociopolitical drivers;
- science and technology drivers;
- cultural and religious drivers; and
- physical, biological, and chemical drivers.

These globally aggregated drivers appear exogenous to decision-makers. Their current condition cannot be influenced effectively. Changes in these drivers are generally slow and are the cumulative effect of many diverse local and regional decisions. But viewed with a longer perspective, these drivers become subject to the influence of human decisions (that is, become endogenous). For example, today's population can be closely estimated and is truly exogenous. Today's decision-makers have no influence over the number of people in the world now. However, national rates of population growth (determined by birth and death rates adjusted for migration) could change substantially because of political decisions—that is, become endogenous—and could influence population half a century hence.

The Decision-maker within the Ecosystem

The influence of humans over ecosystems is most obvious at the local level. People living within an ecosystem undertake myriad activities that alter its condition and capacity to deliver useful services. We highlight important elements of this interaction in Figure 4.1, which is based on the MA conceptual framework diagram. (See Chapter 1.) In the lower left, the ecosystem is represented by the background rectangle. A typical ecosystem has many different decision-makers (farmers, fishers, households, local production communities) with control over some part of the system. We will refer to this unit as an agricultural field in this section for ease of exposition, but it could just as well be a lake, a forest district, or a marine region. The decisions made about the field, and the actions that follow, affect the condition of the ecosystem and the services it provides—both within the field and elsewhere.

The decision-making process is complex and multidimensional. The local decision-maker might be motivated by tradition (my family has farmed

FIGURE 4.1   Decision-making, Drivers, and Ecosystem Services at the Local Level

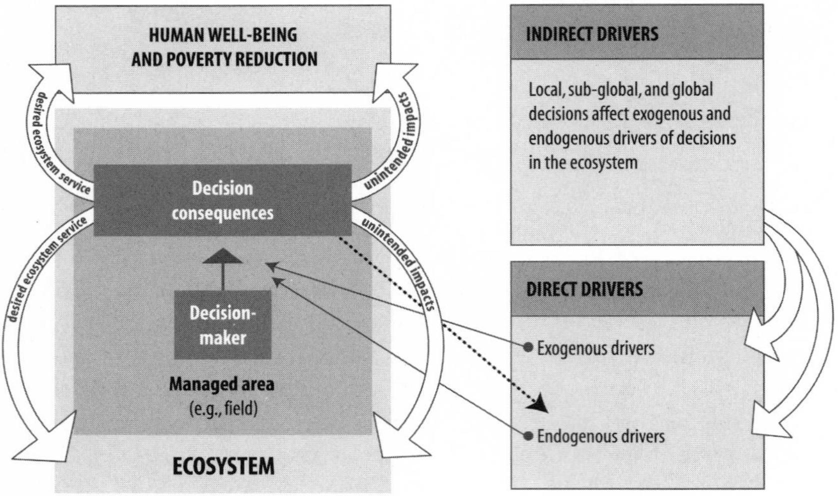

this land for centuries), by biophysical factors (this land and climate is the most productive throughout the year to grow flowers for the international market), by economic need (I sell crops at the local market to buy cloth-ing and medicine), or by familial responsibilities (my children need edu-cation for a better future). The actual decision is based on a combination of many different motives and influences—some are observable, while oth-ers are not.

It is also important to realize that it is the actions arising from the decision that ultimately drive changes in the ecosystem. It is useful to distinguish here between the resulting physical drivers of ecosystem change (direct drivers) and the signals that motivate the decision-making process (indirect drivers). Furthermore, some drivers are under the control of the decision-maker (endogenous drivers) and some are not (exogenous driv-ers). (See Box 4.2.) These categories are indicated by the boxes on the right side of Figure 4.1 and the arrows between the drivers, the ecosystem, and the decision process.

As discussed elsewhere in this chapter, the condition of an ecosystem is influenced by natural drivers, such as climate and biological processes, over which the decision-maker has no control. These direct drivers also condition the decisions made. The natural capacity of an ecosystem is a

---

BOX 4.2   Examples of Exogenous and Endogenous Drivers at the Local Level

Selected exogenous local drivers include:

- indirect drivers that influence the decision-making process
  - institutions (such as property rights, community organizations, or marketing regulations),
  - prices and markets, and
  - technology development.
- direct drivers that directly affect ecosystem condition and services
  - some ecosystem characteristics, and
  - local effects of regional and global environmental change (such as increased mean temperature from rising carbon dioxide concentrations or lower mean temperature from volcanic pollution).

Selected endogenous local drivers include:

- indirect drivers that influence the decision-making process
  - technology adaptation (such as fish location technology or precision agriculture).
- direct drivers that directly affect ecosystem condition and services
  - changes in local land use and land cover,
  - species introductions and removals, and external inputs (such as fertilizer use, pest control, or irrigation water).

---

function of abiotic (nonliving) and biotic (living) characteristics, including geomorphology, soil quality, weather conditions, and biodiversity. The natural capacity makes some potentially desirable ecosystem services biologically impossible (growing coffee in Canada, for instance) while others would require dramatic changes in the ecosystem. The natural capacity differs strongly between localities and regions. Coarse-scale patterns are largely determined by climate and parent material of the soil, while fine-scale patterns are defined by ecological, soil, and management processes and by land use history. The state of the natural capacity at the time a decision is taken sets the initial condition for the range, level, and quality of intended services that can be obtained. In addition, a potentially large number of local, sub-global, and global forces influence the decision process directly and therefore the drivers under control of the decision-maker.

The effect of an externality (indicated by the arrows to the right of the decision consequences in Figure 4.1) is seldom confined to the environs of the decision-maker. External effects extend to other parts of the ecosystem and even to other ecosystems. It is possible for individually unimportant external effects to have dramatic regional and global consequences when many local decision-makers simultaneously take decisions with similar unintended consequences.

Ecosystem Consequences of Decisions Outside an Ecosystem

The discussion thus far has focused on decision-making within an ecosystem and the decision-making process that directly affects the ecosystem and its services. But there are many decisions made to bring about outcomes that are not directed at a specific ecosystem or its services. What we call the regional level is intended to encompass all these other decision-makers.

One way to categorize these decision-makers is the distinction between private (entrepreneurs and business) and public (government and non-governmental organizations [NGOs]). As a generalization (albeit with many exceptions), the private decision-maker has personal gain as a primary motive while public decision-makers are motivated by the well-being of the unit for which they are making decisions. Private decision-makers include individuals and communities that make collective decisions for local, national, and global businesses. Political decision-making takes place in units that include nations, sub-national units (county, district, municipal, province, or state), supra-national units (groups of nations such as the European Union that have some common legal, economic, and political institutions), and trading communities (such as the North American Free Trade Association or regional groupings such as the South Asian Association for Regional Cooperation).

Regardless of the motivation of these decision-makers, few if any of the units for which decisions are made are synonymous with an ecosystem. A county, state, or nation can encompass multiple ecosystems. Or a single ecosystem can cross multiple jurisdictional boundaries. As a result, efforts to mitigate negative externalities often require negotiations among multiple decision-makers with differing interests.

An almost limitless range of interactions is possible between the regional and the local levels. The ecosystem decision-maker at the local level uses inputs from the regional level in the process of enhancing provisioning and supporting ecosystem services. Ecosystem services from the local level, intended or unintended, are inputs into activities at the regional level. In turn, decisions made at the regional level can affect multiple ecosystems. Some regional decisions are intended to influence ecosystem conditions and services. Examples include land, water, and natural resource policies. Many other decisions taken at this level with no intent to affect an ecosystem nonetheless have consequences for it. The intent of these decisions is to influence activities in the domains over which they have some control—a political, business, or community unit—that does not necessarily correspond with an ecosystem or biome.

A wide range of factors influences decisions at the regional level. Unlike local ecosystem decisions, however, many more factors at this level are endogenous. The number of exogenous factors depends in part on the relationship among the various units involved. Although we refer to the region as a single level, it actually encompasses many nested and overlapping levels of control and decision-making. For example, most nations have sub-national political units (states or provinces), and these units are often further divided into counties, districts, and municipalities.

The sub-national, national, international structure provides a natural hierarchy of endogenity for drivers. Decision-making at higher levels influences factors that are exogenous to decision-makers at lower levels. For example, international grain markets collectively determine world wheat prices; national governments can influence prices that farmers receive with trade and production taxes and subsidies, but farmers treat those prices as exogenous. Or a national government can set air pollution standards that affect sulfur dioxide emissions from individual power plants. For the plant manager, the regulation is exogenous, and for the forest managers downwind, the reduction in acid rain is exogenous.

But this hierarchy is by no means exclusive. Some drivers are endogenous at the local level but exogenous at the regional level. For example, land use management rules such as zoning regulations are frequently a local decision over which the state or national government has no control. Furthermore, the degree to which a driver is outside the influence of a decision-making process depends to some extent on the temporal scale. Some factors may be exogenous in the short run but subject to change by a decision-maker over longer periods.

At the regional level, then, the endogenous drivers of decision-making often include:

- institutions (such as property rights or trade barriers),
- service and commodity prices and markets,
- technology development, and
- macroeconomic policy.

The exogenous drivers include:

- changes in land use and cover patterns,
- developments in basic science, and
- ecosystem characteristics.

Drivers of Ecosystem Change

The most important sets of drivers that the MA will use play out at all levels (global, regional, and local), but in differing time frames and combinations.

Demographic Drivers

The demographic variables that have implications for ecological systems include population size and rate of change over time (birth and death rates), age and gender structure of a population, household distribution by size and composition, spatial distribution (urban versus rural and by country and ecosystem), migration patterns, and level of educational attainment.

The interactions among population and ecosystems are complex. Population size and other demographic variables influence the use of food, fiber, clean water, energy, shelter, transport, and a wide range of ecosystem services. Increases in population decrease the per capita availability of both renewable and nonrenewable resources. When coupled with growing income and other factors such as urbanization and market development, population growth increases the demand for food and energy.

Demographic projections suggest that future population growth rates will not be uniform throughout the world. At least 95 percent of the additional 3 billion or so people likely to inhabit the planet in the next 50 years will live in developing countries, and most will be in the tropics and sub-tropics. The U.S. Census Bureau projects a world population of 9.1 billion by 2050 (U.S. Census Bureau 2002), while the median projection of the United Nations Population Bureau for 2050 is 8.9 billion (UN Population Division 2001). Other projections, however, cite both higher and lower numbers. In 1985, 75 percent of the world lived in developing countries; this increased to 78 percent by 2000 and is projected to reach 86 percent by 2050 (UN Population Division 2001). Estimates are that the 49 lowest-income countries, which are mainly in the tropics and sub-tropics, will almost triple their population—from 668 million to 1.86 billion—by 2050 (UN Population Division 2001).

The location of the increases in population has important consequences for ecosystems at the local, regional, and global level. In the past 50 years, for example, on average 90 percent of food was produced in the country of consumption (FAO 2003). If there is no significant change in this ratio, and if the expected population growth in the tropics and sub-tropics materializes, tropical and sub-tropical ecosystems will need to provide significantly more food in addition to the services they already provide. A further complication is that agricultural productivity in the tropics and sub-tropics is projected to suffer from human-induced climate change. Hence these ecosystems will be under considerable pressure in the coming decades. It should also be noted that nearly 50 percent of the current human population live in the 12 megadiversity countries; where the population growth rate is expected to exceed that of the global average, these unique ecosystems will be under significant pressure (Secretariat of the Convention on Biological Diversity 2001).

In contrast to the tropics and sub-tropics, the population of some regions, such as Eastern Europe, is projected to decrease over the next 50 years (U.S. Census Bureau 2002). The implications of negative population growth on economic performance and ecological systems are uncertain.

During the past 30 years, there has been a rapid increase in the percentage of people living in urban centers, a trend that is expected to continue over the next 30 years. In the period 2000–2030, world population is expected to increase by 2.2 billion people, of which 2.1 billion will be urban dwellers. In 1950, 30 percent of the population lived in urban areas; by 2000, the urban population increased to 47 percent, and it is projected to reach 60 percent by 2030 (UN Population Division 2002). In 1975, there were five megacities (with 10 million or more residents)—two in industrial countries and three in developing countries. By 2000, there were 19 megacities, of which 15 were in developing countries. And, by 2015, it is projected there will be 23 megacities, of which 19 will be in developing countries (UNFPA 2002).

Another important demographic dimension is the interaction between population growth and the distribution of income across individuals, countries, and regions. A combination of extreme poverty for many, low national income growth, and weak property rights can, in some instances, greatly increase pressure on fragile, marginal ecosystems. On the other hand, wealthier societies are associated with high consumption patterns of energy and biological resources, which has its own implications for the demand for ecosystem services.

Age, gender, and levels of education are also important demographic variables. Persons with different educational levels tend to vary in their impacts on the environment and in their vulnerability to environmental change. The number and distribution of households by size and composition is important as well. Greenhouse gas emissions, and hence human-induced climate change, can be assumed to depend on the number of households, not just the number of individuals (Roberts and Grimes 1997).

The most crucial population variable in the long run is the rate of change both locally and globally, which is the nexus of birth, death, and migration rates. While the most hopeful dimension of the population equation is that the global growth rate is falling as families around the world choose to have fewer children, a less hopeful sign is that life expectancy has declined dramatically in the Russian Federation due to changes in economic condition and in many sub-Saharan African countries due to HIV/AIDS.

The bottom line is that demographic variables are critical drivers of the demand for ecosystem services and the capacity of the global ecosystem to provide them. Barring major dislocations, such as world war or pandemics, the number of people alive in 2050 and their geographic distribution is an endogenous variable. Decisions made at national and subnational levels can have a dramatic effect on population growth rates through their impact on sociopolitical and cultural factors—in particular, in opportunities for education and the advancement of women and in urban-rural distribution. Decisions at the supra-national level can influence migration across national boundaries.

Economic Drivers

Economic and social well-being are clearly affected by global economic growth and its distribution by country, sector, and individual. How growth is distributed determines the character of demand for ecosystem services. Global economic performance is more than simple changes in national economic activity. International trade, capital flows, and technology are crucial elements in global growth and its consequences for the world's ecosystems. Moreover, the unprecedented rate of global interconnection is leading to dramatic changes in lifestyles and consumption patterns; the consequences of this for global ecosystems are not yet clear.

Global economic trends that began in the last century will likely persist and probably strengthen as the twenty-first century unfolds. First, growth in international trade flows continues to exceed growth in global production, and the differential may be growing. Between 1990 and 1998, for example, the 12 fastest-growing developing countries saw their exports of goods and services increase 14 percent and their output 8 percent (World Bank 2002a). However, not all trade flows are equal in their effects on growth. Dollar and Collier (2001) found that the countries experiencing the most rapid trade-driven economic growth were trading a large share of high technology products. Therefore changes in the volume, value, direction, and composition of trade must be carefully evaluated, along with the degree of restrictions on flows. New and expanded regional and global trade agreements and institutions, such as the World Trade Organization, will likely increase the importance of international trade in global economic performance.

Financial flows and policies affecting international capital movements are also critical. The trend of the late twentieth century toward more open economies led to greater uniformity in macroeconomic (monetary, fiscal, and exchange rate) policies across the world. This trend is manifested in increasing capital mobility and flexible exchange rate regimes, encouraged by institutions such as the International Monetary Fund, the World Bank, and regional development banks. But not all developing countries participated equally. For instance, the vast majority of private-sector capital flows is concentrated in the 10 largest developing countries (World Bank 2002b).

Identifying the key interactions among the rate of growth of an economy, the degree of inequality in the ownership of resources, and trade and capital flows is crucial to understanding their impacts on land use patterns, resource extraction, water diversion and pollution, biodiversity losses, and the landscape. Equally important is understanding the impact of sector-specific subsidies and taxes (on agriculture, energy, and so on), particularly in industrial countries, on local and global ecosystems.

There is some controversy about whether the outcomes of global economic growth are sustainable. There is little question that some of the world's ecosystems have experienced unsustainable pressure. However, the evidence on which this statement is based could be improved considerably. There is a need for a systematic assessment of the potential negative impacts of growth on the resource base in both industrial and developing countries. There is also evidence that the structure of economic growth has an impact on the extent of ecosystem pressure. Demand for services (as opposed to manufacturing), which tend to have fewer negative externalities, rises with income. In addition, as per capita incomes rise, there is greater willingness to pay for mitigation and remediation.

Sociopolitical Drivers

"Sociopolitical" is a word that attempts to capture all the forces that lie in the large conceptual space between economics and culture that affect decision-making at all levels. Indeed, the distinction between sociopolitical and cultural factors blurs as the time scale is extended (Young 2002). Sociopolitical driving forces have been important in the past (e.g., Redman 1999; de Vries and Goudsblom 2002) and should be explicitly included in the MA.

Four categories of sociopolitical forces appear to be undergoing major changes at the beginning of the twenty-first century:

- The general role of the public in decision-making appears to be expanding, as evidenced by the extent of democratization. Despite some backsliding, there is a declining trend in centralized, authoritarian governments and a rise of elected democracies. As well, there is some evidence of improving governance across the developing world.

- The voices that are heard and how they are expressed has changed, as evidenced in the changing role of women and the rise of civil society. Democratic institutions have also encouraged decentralized decision-making, with the intended beneficiaries having a greater say in the decisions made. This trend has helped empower local communities, especially rural women and resource-poor households. Decentralization trends have also had an impact on decisions made by regional and international institutions, with the increasing involvement of NGOs and grassroots organizations, such as traditional peoples groups.

- The mechanisms by which nations solve their disputes, peaceful and otherwise, are changing. Although the cold war has ended, the persistence of regional and civil wars and other international conflicts in some parts of the world continues to be a matter of concern. There is an urgent need to understand the driving forces behind such conflicts and their impact on sustainable livelihoods and the natural resource base.

- The declining importance of the state relative to the private sector—as a supplier of goods and services, as a source of employment, and as a source of innovation—is evident. The future functions of the state in provisioning public goods, security, and regulation are still evolving, particularly in the developing world. In both the developing and the industrial world, the implications of privatization trends on the sustainable management of the local and global resource base are still not clear.

Scientific and Technological Drivers

The development and diffusion of scientific knowledge and technologies can have significant implications for ecological systems and human well-being. Rates of investment in research and development, rates of adoption of new technologies, changes in the productivity and extractive capabilities of new technologies, and the access to and dissemination of information through new technologies all have profound implications.

The twentieth century saw tremendous advances in the understanding of how the world works physically, chemically, biologically, and socially and in the applications of that knowledge to human endeavors. From the introduction of the automobile in the early years to commercialization of genetically modified crops and widespread use of information technology in the later years, many new products drew both praise and damnation regarding their effects on ecosystems. The twenty-first century is likely to see continued breathtaking advances in applications of materials science, molecular biology, and the information revolution—with real potential to improve human well-being around the planet. But these developments have uncertain consequences for ecosystems.

Humans have been extremely successful in institutionalizing the process of scientific and technical change. The organizational structures that encourage researchers to make breakthroughs and use them to develop potentially valuable products—such as research universities, publicly funded research centers, public-private collaborations for research and development, regulatory institutions, and international agreements that collectively determine intellectual property rules—are either in place or being implemented in the industrial world. However, they are not in place in most developing countries. Furthermore, institutions to facilitate use of, and compensation for, indigenous knowledge are not well developed.

Society's ability to manage the process of product dissemination—identifying the potential for adverse consequences and finding ways to minimize them—has not always kept pace. This disparity became especially obvious as the introduction of genetically modified crops met widespread opposition in many parts of the world. The protests in part resulted from the speed of advancement, as the rate of commercial adoptions of the first products of this new technology was unprecedented in a number of countries. At least 30 years passed between the development and widespread use of hybrid maize in industrial countries. For semi-dwarf rice and wheat in developing countries, a similar rate of use was reached only 15 years after development began (Babinard 2001). But use of genetically modified soybeans reached similar levels of use after only 5 years in Argentina and the United States. The use of the Internet accelerated worldwide communication and the organization of protests.

The state of scientific and technical knowledge at any given point in time depends on the accumulation of knowledge over time. Decision-makers can, however, affect the rate of change in scientific and technical knowledge through setting research priorities and changing levels of funding. Domestic government funding for science and technology is driven by objectives such as scientific education, technology development, export markets, commercialization and privatization, and military power. International donors strongly influence science and technology in developing countries, primarily through the type of research they are willing to fund. The private sector responds to the perceived future for their products, looking for those that will be the most acceptable and profitable.

Drivers Determined by Cultural and Religious Values

The word "culture" has many definitions in both the social sciences and in ordinary language. To understand culture as a driver of ecosystem change, it may be most useful to think of culture as the values, beliefs, and norms that a group of people share. In this sense, culture conditions individuals' perceptions of the world, influences what they consider important, and suggests courses of action that are appropriate and inappropriate. And while culture is most often thought of as a characteristic of national or ethnic groups, this definition also acknowledges the emergence of cultures within professions and organizations, along with the possibility that an individual may be able to draw upon or reconcile more than one culture.

There is a substantial literature examining the role of culture in shaping human environmental behavior. It focuses primarily on variations within a nation rather than across nations, in part because it is extremely difficult to establish causal effects of a variable as broad in conceptualization as culture. Two central concerns of the literature are the degree to which the environmentally salient parts of a culture are amenable to change and the degree to which culture actually influences behavior with regard to the environment. There is considerable debate about the first concern. Again, broad generalizations are not warranted, but it is clear that some aspects of culture can change with great rapidity while other elements are inherently conservative.

A substantial body of literature provides lessons on how policies and programs can most effectively produce cultural change around environmental behavior (Dietz and Stern 2002). Obviously, the relationship between culture and behavior is context-specific. Indeed, one important lesson of research on this topic is that overarching generalizations are seldom correct, that the ability of culture to shape behavior depends on the constraints faced by individuals, and that the effects of changing constraints on behavior depend on the culture of the individuals encountering the changes (Gardner and Stern 1995; Guagnano et al. 1995).

At least since it was argued by White (1967) that environmental disruption is a result of some elements of Judeo-Christian culture, there has been special interest in the role of religion in shaping environmental behavior. Arguments that major world religions have led to national or regional differences in environmental impact have not been sustained. However, there is a growing body of scholarship that examines how variations in religious beliefs within a society are related to environmental beliefs and values (Eckberg and Blocker 1989; Kempton et al. 1995; Eckberg and Blocker 1996). In addition, theologians have begun exploring in detail the teachings of the major world religious traditions with regard to the environment. Finally, religious precepts that prescribe acceptable and unacceptable consumption patterns might have a significant impact on the demand for ecosystem services as population grows.

Physical, Biological, and Chemical Drivers

There are natural and human-induced physical, chemical, and biological drivers of change. Natural drivers include solar radiation, climate variability and extreme weather events (such as droughts, floods, hurricanes, and cyclones), fires, volcanic eruptions, earthquakes, pest and disease outbreaks, and natural biological evolution. The primary human-induced drivers include land use changes, climate change, air and water pollution, acid deposition, soil erosion, soil salinization and fertility changes, irrigation, fertilizer use, harvesting, the use of persistent organic chemicals, and the introduction of non-native species.

Key physical and biological characteristics include the living (plants, animals, and microorganisms) and nonliving (atmospheric composition, climate, soil, terrain, rivers, lakes, and oceans) components of the Earth system that sustain ecosystems and human life. Earth has evolved over millions of years through the interactions between living organisms and their environment. These interactions facilitated new life forms and landscapes, and the current conditions of a life-supporting atmosphere.

Human societies have for centuries affected the local environment through land use practices, domestication of plants and animals, and the introduction of exotic species to an area, but the cumulative effect of their activities are now for the first time dominating many regional and global processes—biodiversity, global biogeochemical cycles, and climate (IPCC 2002)—in part driven by increasing demand for food, fiber, clean water, energy, minerals, and transport. Understanding how human activity affects the basic geology and biology of the planet is crucial to assessing the future capacity of the global ecosystem.

Many of these drivers are changing and are projected to continue to change in the coming decades in many parts of the world, as indicated by these examples:

- conversion and fragmentation of ecosystems in many parts of the world, as illustrated by an annual rate of tropical deforestation of about 0.7 percent (Houghton et al. 2001);

- climate change, with the expectation of warmer temperatures, changes in precipitation, and increases in extreme weather events such as heat waves, floods, and droughts and associated fires and pest outbreaks (Houghton et al. 2001; McCarthy et al. 2001);

- a global rise in sea level (Houghton et al. 2001; McCarthy et al. 2001);

- degradation of air, water, and land, especially in many developing countries (Stafford-Smith and Reynolds 2002); and

- planned and inadvertent introductions of nonnative species (Heywood and Watson 1995; Dukes and Mooney 1999).

Interactions among Drivers

Changes in ecosystem services are always caused by multiple, interacting drivers originating from different levels of organization of the coupled human-environment systems. For example, many changes are driven by a combination of drivers that work over time (such as population growth and climate change) and drivers that happen intermittently (droughts, wars, or economic crises, for example). There are functional interdependencies between the drivers of changes in ecosystem services, both at each organizational level (horizontal interplay) and between levels of organization (vertical interplay) (Young 2002).

Moreover, the changes in ecosystem services lead to feedbacks on the drivers of changes. For example, changes in ecosystems create new opportunities and constraints for land use, induce institutional changes from local to global levels in response to perceived and anticipated resource degradation, and give rise to social changes in the form of income differentiation (as there are winners and losers in environmental change).

The drivers of change may follow different modes of interactions:

- One cause may temporarily dominate other drivers in a certain period. For example, local changes in ecosystems are caused not by climate change but by habitat loss. This fact has been used by non-biologists to argue that climate change is of little importance to ecosystems. This approach, however, effectively ignores small, systematic trends in drivers that may become important in the longer term (Parmesan and Yohe 2003).

- Factors driving changes in ecosystem services can be connected as causal chains—that is, interconnected in a way that one or more variables (indirect drivers, mainly) drive one or more other variables (direct drivers).

- Different factors can intervene at the same time—for instance, independent but synchronous operation of individual factors can lead to land change.

- Different factors may also intervene in synergetic factor combinations—that is, several mutually interacting variables drive changes in ecosystem services over time.

Reviews of case studies reveal that the most common type of interaction is synergetic factor combinations (Geist and Lambin 2002). This implies combined action of multiple drivers that produces an enhanced or increased effect due to reciprocal action and feedbacks between drivers.

The complexity in the interactions among drivers of changes in ecosystem services can be greatly reduced by recognizing that there are a limited number of ways in which these drivers are actually combined. For any given human-environment system, a restricted set of drivers is essential in order to predict the general trend in the ecosystem. This makes the problem tractable. This idea is the basis, for example, of the syndrome approach (Petschel-Held et al. 1999), for the analyses of trajectories of environmental criticality (Kasperson et al. 1995), of major spirals of household impoverishment and environmental degradation (Kates and Haarmann 1992), of pathways of land use change (Lambin et al. 2001), and of spatial economic models of land use change (Nelson and Geoghegan 2002).

Models have captured some of the generalizable patterns of change that result from recurrent interactions among driving forces. For example, the environmental Kuznets curve describes the relationship between environmental degradation and economic growth, which holds true for a range of ecological issues—those at the local scale, which affect a population in the short term (Kuznets 1979). Case studies also identify specific sequences of events leading to changes in ecosystem services. Tropical deforestation sometimes results from a sequence of extraction of timber plus initial colonization, for instance, followed by the establishment of colonists with greater access to capital. Competition for access to land takes place and leads to increasing land holdings for the winners, while the losers are pushed to expand the agricultural frontier further. If cattle provide the largest economic rewards for the winners, given market conditions and government subsidies, large-scale land conversion to pasture follows. This, in turn, drives up land prices, leading to further land consolidation (Lambin et al. 2001). In other cases, macroeconomic decline generates large numbers of unemployed people who move into forest areas that are effectively open access. They survive by clearing forest patches of subsistence crops and converting wood to charcoal for sale (Cruz and Repetto 1992). Even though these sequences may play out differently at the detailed level in specific situations, their identification may confer some predictive power by analogy with similar pathways in comparable regional and historical contexts.

The many processes of globalization lead to new forms of interactions among people and between drivers of changes in ecosystem services; they amplify or attenuate the driving forces by removing regional barriers, weakening national connections, and increasing the interdependency among people and between nations. Globalization can either accelerate or buffer the impact of sectoral drivers on ecosystems, but it always gives rise to a greater level of functional interdependencies among drivers between local, national, regional, and global levels.

5 *Dealing with Scale*

EXECUTIVE SUMMARY

- There is seldom one, ideal scale at which to conduct an ecosystem assessment that will suit several purposes. The Millennium Ecosystem Assessment (MA) advocates a multiscale approach.

- Many environmental problems originate from the mismatch between the scale at which ecological processes occur and the scale at which decisions on them are made. Outcomes at a given scale are often critically influenced by interactions of ecological, socioeconomic, and political factors from other scales. Focusing solely on a single scale is likely to miss such interactions, which are critically important in understanding ecosystem determinants and their implications for human well-being.

- The choice of scale and boundaries of an assessment is not politically neutral. It can implicitly favor certain groups, systems of knowledge, types of information, and modes of expression. Reflecting on the political consequences of scale and boundary choices is an important prerequisite to exploring how multiscale and cross-scale analysis in the MA might contribute to decision-making and public policy processes at various levels.

- Ecosystem processes and the services they deliver are typically most strongly expressed, most easily observed, or have their dominant drivers or consequences at particular scales in space and time. The spatial and temporal scales are often closely related, defining the scale domain of the process.

- Social, political, and economic processes can be more readily observed at some scales than others, and these may vary widely in terms of duration and extent. Furthermore, social organization has more- or less-discrete levels, such as the household, community, and nation, that correspond broadly to particular scale domains in time and space.

- Assessments need to be conducted within a scale domain appropriate to the processes or phenomena being examined. Those applicable to large areas generally use data at coarse resolutions, which may not detect fine-resolution processes. Even if data are collected at a fine level of detail, presentation of the findings at a larger scale means local patterns, anomalies, and the exceeding of thresholds disappear.

- A multiscale approach that simultaneously uses larger- and smaller-scale assessments can help identify important dynamics of the system that might otherwise be overlooked. Trends that occur at much larger scales, although expressed locally, may go unnoticed in purely local-scale assessments.

■ If an assessment covers a shorter time period than the time scale of important processes, it will not adequately capture variability associated with, for instance, long-term cycles such as climatic or economic trends. Slow changes are often harder to detect than rapid changes, given the short period for which data are available.

■ Given the pervasive influence of scale on any conclusions reached, it is essential that assessments be explicit regarding the geographic extent and period of time for which the study is valid. The same applies for data sets that are used in assessments.

Introduction

Scale refers to the physical dimensions, in either space or time, of phenomena or observations (O'Neill and King 1998). This is expressed in physical units, such as meters or years. In the Millennium Ecosystem Assessment (MA), the word "level" is used to describe the discrete levels of social organization, such as individuals, households, communities, or nations (Gibson et al. 2000). A level of organization is not a scale, but it can have a scale (Allen 1998; O'Neill and King 1998).

It is necessary to distinguish the "scale of observation" from the "scale of the phenomenon." The scale of observation is a construct based on human systems of measurement. Observation scale has three components: extent (or duration), resolution, and grain (Blöschl and Sivapalan 1995; Blöschl 1996). The extent is the total area or time over which a phenomenon is observed, the resolution is the interval or distance between observations, and the grain is the area or duration of an individual observation. These concepts are illustrated in Figure 5.1. Independent of the scales at which things are observed by humans and their instruments, there are characteristic scales at which both ecological and human processes occur. The characteristic scale of a process describes the typical extent or duration over which the process is expressed—that is, over which it has its impact. The scale domain of a process is defined in terms of both its characteristic space and time scales. The grain of a phenomenon is a concept distinct from the grain of observation, and refers to the smallest unit that is internally homogenous.

In the MA, unless otherwise stated, the word "scale" means the extent or duration of observation, analysis, or process. For instance, an assessment can be said to be "at the regional scale," or the time scale of the El Niño phenomenon is "at the decadal scale." The term "large scale" indicates something of greater extent than "small scale." This conforms to the

---

FIGURE 5.1   Three Components of Scale of Observation

The scale of observation can be described in terms of its extent, resolution, and grain. For example, in observing river discharge over time, grain refers to the time spent taking each sample, resolution refers to the time between observations, and extent refers to the total time period over which the samples were taken (based on Blöschl 1996). Similarly, in observing household expenditure in a particular area, grain refers to an individual household, resolution refers to the density and distribution of observed households over space, and extent refers the total area over which observations were made. In the special cases of continuous digital images or data recorders, grain is equal to resolution and in the former case is referred to as a pixel. Grain can also refer to the characteristics of the phenomenon itself—the smallest unit that is internally homogeneous, independent of the observer.

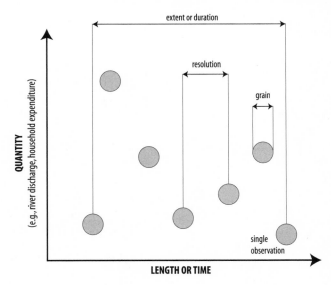

natural language usage of those phrases, although it is the opposite of common usage in cartography. "Long term" and "short term" are used in a relative sense for the time dimension, while "higher level" and "lower level" are used to reflect hierarchical institutional or social organizational levels. Higher levels have greater scope of operation or influence than lower levels.

Emergent properties may appear at some scales or levels of organization. The description of these properties is confined to those particular scales. An emergent property is a phenomenon that is not evident in the constituent parts of a system but appears when they interact as a system. For example, the cultural or recreational value of an ecosystem is often an emergent property at the scale of a landscape (a heterogeneous area consisting of a mixture of different patches, typically many kilometers in extent). There is debate about whether emergent properties have objective reality or are just a useful way of understanding and describing a system (Giampetro in press).

Scale is also related to variation and predictability: small-scale events show more variability than large-scale events do. This is because the effects of local heterogeneity are averaged out at broader scales, so that patterns appear to be more predictable (Weins 1989; Levin 1992). Conversely, models or assessments focusing on broad-scale patterns lose predictive accuracy at specific points in space and time (Costanza and Maxwell 1994).

Why Scale Matters

In assessing ecosystem services, scale matters for two main reasons. First, ecological and social systems and processes operate at a wide variety of scales—from very small and short to very large and long—and between scales they can change in their nature and sensitivity to various driving forces. It cannot be assumed that results obtained at one scale are automatically valid at another (Kremen et al. 2000; McConnell 2002).

Thus if the impacts of processes are observed or assessed at scales significantly smaller or larger than their characteristic scale, there is a likelihood of drawing the wrong conclusion. For instance, it is inappropriate to draw any conclusions regarding long-term trends based on short-duration time series data. People do not infer that the primary productivity of the world is declining just because in the Northern Hemisphere leaves die in the autumn; based on experience, it is obvious that this is part of a longer-term seasonal cycle. Nor can it be assumed that because a change is occurring at one location it is occurring equally at all locations. It is springtime in the Southern Hemisphere during autumn in the North.

Second, cross-scale interactions exert a crucial influence on outcomes at a given scale. Focusing solely on a single scale can miss these interactions. Looking at a particular issue top-down, from the perspective of larger scales or higher institutional levels, can lead to different conclusions than looking at the same issue bottom-up, from the perspective of smaller scales or lower levels (Berkes 2002; Lovell et al. 2002). The scale of the assessment influences both the framing of an issue and the range of possible actions and institutional responses. Where cross-scale interactions in ecological and social systems occur, there should be no expectation of finding a single most appropriate level for response or policy. In most cases, mutually supportive policy changes and responses at different levels are required in order to bring about desired results.

Individual systems important to ecological and social change arise from different scale domains of nature and society (Clark 1985; Peterson and Parker 1998). Scale has close connections with where, how, and by whom decisions are made regarding the use of ecosystem services. Scale also relates to how different actors learn about ecological change. Integrated understandings are of necessity nearly always place-based, and aggregates of observations drawn from smaller scales nearly always contain more detail than observations at a very general scale, offering opportunities for richer kinds of learning (e.g., Environment Canada 1997).

The "domain argument" (Wilbanks and Kates 1999) for conducting a multiscale assessment becomes even more persuasive when taking into account the many interactions across scales:

- Human rules and behavioral norms are embedded in scale-dependent institutional structures such as the boundaries of political jurisdictions. The spatial range of individual actions is restricted to the area of access rights: an agricultural plot, a forest patch, or a lake, for instance. Yet the social, economic, and political structures in which the actors are embedded are of larger scale—provincial, national, or even global. A local assessment is "local" not because it considers only local constraints and processes but because even while it takes account of factors and determinants from different scales, it is framed from the point of view of local stakeholders and it considers decisions and actions taken at that level. To be effective, local assessments must adequately reflect relevant factors and determinants from larger scales.

- Characteristic scales of ecological and human processes often do not match. Thus an integrated assessment of human-ecosystem interactions has to synthesize across scales (Rotmans and Rothman in press). For instance, the characteristic time scale for effective adaptation in ecosystem management depends on both human capabilities for changing management practices and the processes of structural change in the ecosystem.

- Cross-scale interactions can reveal hierarchical systems—that is, systems that "are analyzable into successive sets of subsystems" (Simon 1962:468). Hierarchical systems display a special type of nested orderliness and have particular resilience properties (Peterson 2000).

The minimum requirement for an assessment of ecosystem services is that it should be explicit about the scale and resolution of analysis. The

---

BOX 5.1   The Rationale for Multiscale Assessments

The Millennium Ecosystem Assessment is a multiscale assessment. The main reasons for introducing multiple scales in an already complicated assessment are the following:

- They permit individual ecological and social processes to be assessed at the scale at which they operate and to be linked to processes at different scales and levels of social organization.
- They allow progressively greater spatial, temporal, or causal detail to be considered as the scale becomes finer.
- They allow for independent validation of larger-scale conclusions by smaller-scale studies and create a context at larger scales for findings at smaller scales.
- They permit reporting and response options to match the scales at which social decision-making occurs, with which people can relate, and on which they can act—the local community, the province, the nation, the regional bloc, and the planet.

MA aspires to go beyond this by considering interactions across scales and levels of social organization. From the outset, the MA has been conceived as a multiscale effort, with attention to processes at several scales in space and time and at various institutional levels. (See Box 5.1.)

Changing Scales

Observations drawn from studies at widely different scales are only comparable with considerable care. Comparisons are valid only after careful testing to ensure that scale dependencies have been accounted for. Variables used to describe ecosystem services and their drivers can be thought of as belonging to one of three scaling categories: scale-independent, scale-dependent with known scaling rules, and non-scalable.

Scale-independent variables exhibit conservation of mass or value and show no (or weak) spatial or temporal interdependencies. To make the numerical values of such variables scale-independent, they can be divided by the measurement area (such as per square meter) or duration (per day, for instance). Population density (people per unit area) is an example. Scale-independent variables can be "scaled"—that is, translated from the scale at which the data were collected to a larger or smaller scale—in a very straightforward way through simple addition or proportionality. An example is biomass: the biomass of a hectare of forest is the simple sum of the biomass within each square meter of the forest. If the biomass is homogeneous over the scale of extrapolation, then every piece of the forest does not need to be measured to get the total. It is simply the total area multiplied by the biomass per unit area in the sub-sample.

In the second category, scale-dependent but with known (or potentially knowable) scaling rules, variables are "scalable"—that is, they can be expressed in smaller or larger aggregated units. But they must first be translated to a consistent scale, and the scaling rules may be complex and are often nonlinear. Transpiration is an example. Transpiration from a hectare of forest is not simply the transpiration measured at the scale of a leaf multiplied by the number of leaves in a hectare. This is because the transpiration from one leaf alters the humidity surrounding the leaves downwind of it, and thus their transpiration rate. Scaling of evapotranspiration can be achieved using an explicit model involving a nonlinear coupling constant (Jarvis and McNaughton 1986). Many social and ecological processes belong in this category. They tend to follow nonlinear or discontinuous scaling rules for a variety of reasons, including spatial or temporal interactions (especially feedbacks), organizational scope and the limits of institutional authority, and high heterogeneity or changes in the nature of the regulating factors as the scale changes.

Terrestrial carbon balance is an example of a variable that can be expressed in consistent physical units (grams of carbon per square meter per year) at all scales, but its interpretation changes with temporal and spatial scale. At the time scale of a few minutes and the scale of a leaf, the balance is called net photosynthesis (during the day) or respiration (at night). At the time scale of 24 hours or more, it is called net primary production (if considering plants only) or net ecosystem exchange (plants plus animals plus microbes). Over a period of decades or centuries, rare but large fluxes due to disturbance are included (such as fire, storms, harvest, or pest outbreaks), and the balance is called net biome production. The numerical value of net biome production is one hundredth or less the value of net photosynthesis.

Non-scalable variables or processes are those whose meaning is defined only at a particular scale. The process of decision-making within a household, for instance, may not be scaled up to the nation: different principles apply. Such variables can only be "qualitatively scaled" by placing them in clusters with conceptually related variables at different scales.

Assessments frequently compare or combine observations drawn from studies whose scales were determined independently. The main alternatives for doing so are to convert the observations to a single scale or to seek a multiscale or meta-scale synthesis. Conceptually, it has been suggested that convergence approaches (that is, bringing everything to a common scale) imply that process representations can be transferred seamlessly across scales, while multiscale approaches imply a rejection of that point of view (Bauer et al. 1999).

Converting scale-related information to a common metric often focuses on an intermediate scale, which calls for "downscaling" data about global processes and "upscaling" data about local processes (Wilbanks in press). Scaling up can be as intricate and fundamental a problem as downscaling. Harvey (2000) distinguishes between the use of lumped models, deterministically or statistically distributed models, and models with explicit spatial integration: lumped models use the same model representation for each scale, sometimes with implicit scaling in the form of parameter changes (Bugmann et al. 2000); distributed models use the same equation on a spatially explicit grid; and explicit integration seeks to formulate a correct representation for the higher-scale processes. Scaling up is essentially an aggregation challenge, complicated by the fact that simply adding smaller-scale values can give misleading results. For instance, the data may not meet standards for valid sampling, or they may fail to capture stochastic (random) variability in processes. The challenges are especially acute when larger-scale values are being estimated from incomplete local evidence. A number of technical alternatives for dealing with statistical problems in upscaling have been summarized (Rastetter et al. 1992; Harvey 1997).

Downscaling is a different challenge, involving collecting or estimating data at finer scales (such as regional or local values) from processes studied at larger scale. Modelers use both numerical (model-based) and empirical (data-based) approaches for this (Bass and Brook 1997; Easterling et al. 2000). Problems include limited data availability at detailed scales (and the costs of filling any gaps) and increasing complexity of causal relationships in more integrated, small-scale models. One of the forces encouraging downscaling is the need to provide information relevant to public participation, decision-making, and action at a relatively local scale.

Box 5.2 provides some practical guidance on dealing with scale-related data issues.

Space and Time Domains

One of the grand queries of science is understanding relationships between macro-scale and micro-scale phenomena and processes, between short-

---

BOX 5.2   Suggestions for Working with Information at Different Scales

If possible, convert studies of scale-dependent variables to a compatible scale before comparing or combining them. If not, interpret the studies independently at their individual scales.

Processes with nonlinear dynamics are seldom scale-independent:

- Weakly nonlinear processes may be approximated over limited parts of their range of scales by linear interpolation.

- For strongly nonlinear processes, a biased larger-scale estimate will be calculated if the inputs are averaged and then passed through the process. The correct approach is to calculate the output at each point for which input data are available, and then sum over space or time.

In upscaling, to generate an unbiased aggregation of a sparsely sampled variable from an uneven environment, use a richly sampled indicator that covaries with the variable of interest (a scalar) to create a weighted average.

In downscaling, a probabilistic spatially or temporally explicit disaggregation of a heterogeneous variable can be constructed using a scalar.

It is not always necessary to drop to the finest resolution or to rise to the highest possible level of integration in order to represent cross-scale interactions adequately. To determine or illustrate the cause of a phenomenon, drop down to the next logical scale; to determine the constraints on a process, move up to the next logical scale. This is a rule of thumb, not a rigorous result, but is a useful way to limit the scope of an analysis.

---

term and long-term effects, and during implementation at different organizational levels (Wilbanks and Kates 1999; Gibson et al. 2000; Kates et al. 2003).

The time and space scales of a process are frequently correlated and are together referred to as the scale domain of the process (Bisonette 1997). "Big" processes are often "slow," and "small" processes are often "fast." (See Figure 5.2.) A fast process (or variable) is one that changes rapidly in relation to the life span of the organisms or entities that it is acting on. A slow process changes only gradually relative to the internal dynamics of the system being analyzed. In a forest ecosystem, for example, small and fast scales are dominated by biophysical processes controlling individual plant physiology and morphology. At the scale of a patch (tens of meters), interspecific competition for nutrients, light, and water influence growth, species composition, and succession over a period of decades. At forest stand scales, consisting of many patches, disturbances such as fire and insect outbreaks determine landscape heterogeneity over centuries. At the

FIGURE 5.2   Characteristic Scales in Time and Space for Some Ecological and Social Processes

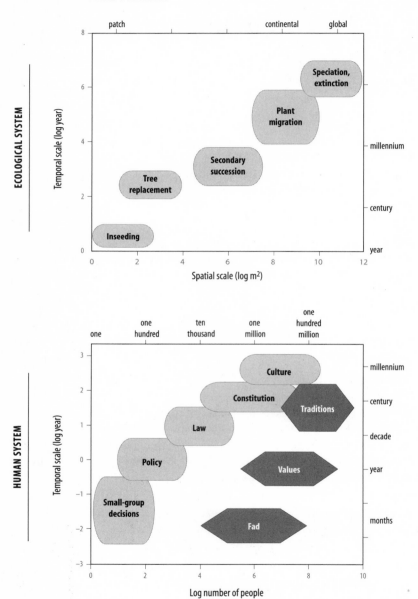

Source: Modified from Delcourt et al. 1983; Gunderson et al. 1995a.

largest scales, climate processes alter structure and dynamics across hundreds of kilometers over thousands of years.

There are analogous space-time domains in social systems. For example, adaptive responses and behavioral changes at the individual level take place within an individual's lifetime, while societal responses often occurs over generations. In neither social nor ecological systems, however, is there necessarily a relationship between scales in space and time: some widespread changes have occurred very rapidly, and some local processes can be slow to change. The correlation between space and time scales can be particularly weak or absent in many modern social processes. It is speculated that this is due to the reach and speed of modern transport and information systems (Goodchild and Quattrochi 1997).

The correlation between large-scale ecosystem changes and long time frames presents a dilemma for ecosystem service assessments. To answer questions about the maintenance or resilience of these services in the future, long-term processes and their interaction with behavior on shorter time scales need to be understood. An implication is that global scale assessments, in particular, may need to consider historical and prehistorical data in order to gain the deeper time perspective necessary for a robust understanding of some large-scale processes. Furthermore, ecosystem assessments should strive to establish baselines against which future changes can be measured.

The links between characteristic scales in time and space can be used, cautiously, to infer long-term consequences by examining patterns over a large space domain. This is called a space for time substitution. For example, because large scales are likely to include areas undergoing rare events, instead of measuring net biome production over long periods (which is at any rate often impossible, given the short historical record), measurements can be done over large areas. Another example is the determination of long-term fire frequency. Many researchers have assumed that fire frequency is equal to the area fraction of the landscape burned per year. This will be untrue if it is the same parts of the landscape that burn repeatedly.

Inertia in Human and Ecological Systems

Both human and ecological systems frequently exhibit a property analogous to inertia in physical systems: the tendency to continue along a pathway of change after the pressure driving that change has been removed. The reason is that many of the processes involved have long time delays built into them. For instance, a fishery catch may continue to rise for a period after the point of sustainable catch has been exceeded, simply because of the maturation of juvenile fish that were hatched before the sustainability limit was passed (Rothschild 1986). Another example is sea level rise in response to climate change: this will continue for centuries after the emissions of greenhouse gases to the atmosphere have been radically reduced (IPCC 2002).

Inertia in the ecological system tends to mute the signals of impending problems and can lead to a tendency to overshoot the target once corrective action has been instituted. The inertia in human systems can cause the implementation of effective action to lag behind the first detection of the problem by years to decades. The combination of these two forms of inertia in coupled ecological-human systems has the potential to allow the system to transgress thresholds that are either ecologically unsustainable or socially unacceptable, and the resultant changes may be irreversible in realistic time frames.

There is a hypothesis, as yet unproven, that the "slow variables" (those with the largest inertia), rather than the "fast variables," are responsible for the resilience properties of a system (see papers in Gunderson and Holling 2002).

Viewing a Particular Scale in Context

Processes that operate at a particular scale are typically related to processes at other scales as well. One familiar example is land use at a local scale, which results from local institutions and actions but is shaped by national policy frameworks and global economic markets. At the same time, local actions may add up either as cumulative changes (such as species extinction) or as systemic changes at a larger scale (such as the effects of emissions of ozone-depleting gases on the stratosphere) (Turner II et al. 1990).

Since pieces of a geographic mosaic are nested within larger pieces, and those are within still larger pieces, it is often useful to think of geographic areas in terms of hierarchies of places and place-related processes. Such approaches lend themselves to the application of hierarchy theory (e.g., O'Neill 1988). Simon (1974) argued that semi-autonomous levels are formed from the interactions among a set of variables that share similar speeds and spatial domains.

In understanding social scales, it may be useful to consider the different forms of hierarchy. In inclusive hierarchies, groups of processes or objects lower in the hierarchy are contained within groups ranked as higher in the systems (such as modern taxonomic classifications). In exclusive hierarchies, groups of objects or processes that are ranked as lower are not contained within the groups at a higher level (military ranking systems, for example). And in a constitutive hierarchy, groups and processes are combined into new units with their own functions and emergent properties (a stakeholder committee, for instance).

Some important social processes do not neatly fit into this concept of nested hierarchies, with space and time strongly correlated. Social networks can introduce very strong connections between places, resulting in interaction across spatial and institutional boundaries. An example is the flow of ideas and coordination of action in different countries through transnational civil society organizations that do not necessarily flow through successive layers of a nested hierarchy. For instance, the Chipko (tree hugger) movement in India was a local-scale action that was quickly internationalized and that inspired similar social environmental movements worldwide. Such flows of ideas or advocacy for coordinated action also tend to be opportunistic, jumping over or skipping scales as they "shop" for the scale or forum that would provide the greatest chance of successful outcomes (Keck and Sikkink 1999).

Transfer of technologies and investments or sharing of stages of the commodity production chain through parts of a transnational corporation are other examples of crossing boundaries. Processes of diffusion of technological and institutional innovations are often critical factors in the use of ecosystem services. Thus network-related concepts are likely to be important in the MA for understanding connections between proximate and primary determinants or causal factors and for identifying possible response options. Network-related concepts are important as well in considering response options at small scales that may be replicated in larger domains without passing through the neat nested hierarchies of governance structures.

Scales in Ecological and Human Systems

The characteristic spatial scales of ecological systems are influenced by numerous factors, including the home range of individual mobile organisms or the range of influence of non-mobile organisms, the geographic distribution of a population of interbreeding organisms, the area over which a disturbance occurs, and the distance over which material is transported during the period for which it is ecologically active. For instance, the effective lifetime of carbon dioxide in the atmosphere is several centuries—during this time it can be transported all over the world. Hence its characteristic scale is global. In contrast, tropospheric ozone can only be transported by wind currents over a few hundred kilometers before it is consumed by atmospheric reactions; thus its characteristic scale is regional.

Characteristic temporal scales of ecological systems are influenced by the life span of organisms, the turnover rate of material pools, and the average period between disturbances at a location. An important distinction, particularly for determining system resilience, is between fast and slow variables or processes. Thresholds of irreversibility are typically related to changes in the slow variables (Gunderson and Holling 2002).

Spatial scales of social, political, and economic processes or variables are shaped by the area of operation, influence, or access rights exercised by various levels of institutions or social organization. Socioeconomic time scales are determined by the response times of humans and their institutions; they may be very rapid (electronic trading of commodities) or relatively slow (institutional change, typically). For example, the characteristic scale of an individual household in a freehold tenure system may be the area of land that the people own; for a community, it may be a village or municipal boundary; and for a country, it is the area included in the national borders and the exclusive economic zone in the ocean. The spatial scales of economic processes typically have a political dimension and are determined by the area over which goods or services are traded, extracted, transported, or disposed of. Economic and political processes are, in turn, embedded within and permeated by sociocultural processes that operate at different institutional levels.

Direct interactions between humans and ecosystems—in agriculture, for example, or forestry or land use—mostly occur at local or micro scales and often at lower institutional levels. (See Figure 5.3.) This can also apply to indirect actions. For instance, although climate change is a global phenomenon, the responses of ecosystems are determined by the changes in the local climate rather than the global average change. Moreover, direct mitigation measures and behavioral responses also typically occur at the local level.

Ecosystem services, though often meeting needs expressed over large scales such as nations, are generally actually delivered at the local scale. Thus an assessment of ecosystem services and their implications for human well-being at global or regional scales typically needs to:

- scale up the ecosystem, taking each service in turn using specific scaling rules, including competition between different actors and different services; and

FIGURE 5.3   Overview of Some Commonly Used Institutional Levels and Ecological Scales

Levels are arranged on a shared vertical axis representing spatial extent. The arrows represent key influences. Direct interactions mostly take place at the local scale, but governance occurs at many scales.

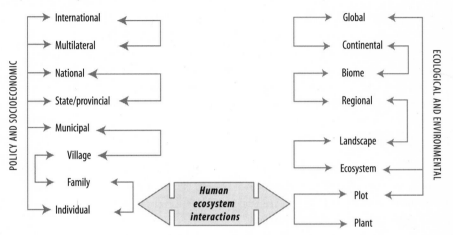

Source: Courtesy of Rik Leemans.

- scale down impacts on ecosystems by scaling down either the environmental pressures (such as by regionalizing the estimates of global climate change) or the socioeconomic activities (by predicting, for example, where a logging company will harvest trees within a large concession area).

It is possible to assess ecosystem services and human well-being more readily at some spatial scales than others. For the MA, as an integrated assessment, these scales are determined by the characteristic scales of both ecological and socioeconomic processes. Thus the planet, the region (a fraction of Earth's surface corresponding to a biome or a major politico-economic bloc), the drainage basin, and the local community tend to emerge repeatedly as chosen scales for the MA.

Temporal scale is an issue in understanding human societies and institutions in a variety of ways. Perhaps the most familiar case is benefit-cost approaches to estimating the value of economic investments at different time scales. For example, in considering investment choices, how should long-term payoffs be evaluated relative to short-term payoffs? The con-

ventional approach is to use a discount rate, which compares future returns from the investment with what could be earned from a neutral investment such as a mutual fund. Use of a high discount rate reduces the
estimated value of longer-term returns, however, and is therefore biased in
favor of investments that yield returns in the immediate or short term.
(See also Chapter 6.)

Temporal scale is also a key consideration in historical studies, including studies of institutional, technological, and sociopolitical change. It is
common to treat short-term, mid-term, and long-term forecasts differently
in terms of methods and assumptions, because the longer term raises so
many uncertainties about contextual assumptions. Quantitative forecasts
of economic and demographic change are usually limited to time horizons
of 25–30 years, if that; the longer term is considered the province of futur-
ists rather than forecasters and is couched in terms of scenarios rather
than predictions.

Most analyses of human systems include multiple time frames. For instance, national politics has a short-term rhythm set by the schedule of
elections, while longer-term trends move in such directions as privatization,
devolution, and democratization.

One subject of research has been whether long-term changes in human systems show regular, predictable fluctuations. As an example, there
may be a Kondratiev cycle of 50–60 years in macroeconomies, perhaps
related to waves of technological change, and Kuznets cycles of 15–20
years related to infrastructure development (Berry 1991). Recent studies
point to possible fundamental relationships between geophysical rhythms
and economic fluctuations (Berry 2000). Much of this work grows out of
efforts to understand implications of natural climate variability over periods of millennia, but it also addresses issues of seasonal, annual, decadal,
and century-scale environmental change.

Scale and Policy

Politics of Scale

The choice of scale is not politically neutral, because the selection may
intentionally or unintentionally privilege certain groups. The adoption of
a particular scale of assessment limits the types of problems that can be
addressed, the modes of explanations that are allowed, and the generalizations that are likely to be used in analysis. This applies to temporal and
spatial scales as well as institutional levels.

For example, the range of ecosystem services that are directly used and acknowledged as having important support functions is dependent on sociocultural contexts, and these are restricted in space. As the assessment is conducted at progressively larger scales, the number of ecosystem services that are fully shared among places, and thus can be mapped "wall-to-wall," drops. The local services that would be visible in a local assessment may no longer be visible in a regional or global assessment. The same basic ecosystem processes (such as net primary production) can be seen as providing different services at different scales—timber at the local scale, but carbon sequestration at the global scale. These issues are critical for the MA because trade-offs between the services are likely. At various scales we need to ask: ecosystem services for whom?

Analyzing these trade-offs requires understanding politics and markets. Many such cross-scale trade-offs are not perceived as such, but instead become conflicts or crises created by a more powerful group (often the state) around the provision of one favored form of ecosystem service from which they can obtain rent or other benefits. Scale can be an argument that empowers state institutions. Most states view indigenous knowledge and institutions as local in scope, relevance, and power, whereas the rules and knowledge of the state are viewed as bigger in scale, scope, and significance. As a consequence of this thinking, there is a strong tendency to override, minimize, or ignore local considerations, issues, or preferences. Many ecosystem management problems result from centralization and uniformity in bureaucratic operations that hinder local adaptation and learning. On the other hand, local adaptation is not universally good. Sometimes a state is needed to deal with the externalities that may arise from local decisions or to arbitrate among competing local claimants to ecosystem services. Scale is thus critical for issues of governance of ecosystems, as discussed in the next section.

Choice of time scales is equally important. If an assessment is focused on short-term concerns, then "important" ecosystem services are deemed to be those that are already or about to be threatened, such as freshwater resources for drinking, fuelwood supplies, or food production. On the other hand, if the users are more concerned with decisions that may have consequences over several decades or centuries, then issues of alterations of carbon balance or opportunity and resilience costs of biodiversity loss become much more important.

Adoption of a global scale immediately places issues such as climate change and carbon management at a much higher priority than, say, sanitation or access to clean drinking water. The attractiveness of the multiscale approach in the MA is that it provides the chance to think about problem identification and response options at more than one scale. It also allows for the analysis of spatial synergies and trade-offs among possible responses.

Likewise, the choice of boundaries is not neutral but has political consequences. For example, setting boundaries as a watershed basin as opposed to a geopolitical identity may make good sense from an ecological perspective, but it may be irrelevant for management if there is no political mechanism to deal with transboundary issues.

Reflection on the political consequences of scale and boundary choices is an important prerequisite to exploring what multiscale and cross-scale analysis in the MA might contribute to decision-making and public policy processes. Designating the boundaries is best done through collaboration between scientists, decision-makers, and representatives from different stakeholder groups.

Institutional Fit and Interplay

Many problems arise from the failure to recognize cross-scale interaction in both ecological and social systems (Young 1994). The effectiveness of institutions governing the management and use of ecosystem services depends not only on their own characteristics but also on how they interact with other institutions. An important class of interactions is those that occur vertically across levels of governance; these often correspond to changes in spatial scale.

The commonest forms of institutional interplay are those between state and local bodies. An emerging arena is the interplay between national institutions and new institutions at regional and international levels (Young 2002). These add a further layer of complexity to both future driving forces of change as well as possible sources of rules and guidance for human choice.

If cross-scale interactions in ecological and social systems affecting ecosystem services are common, then it should not be expected that there is generally a single most appropriate level for response or policy. While responses at certain levels or scales can have disproportionately greater significance or impact, appropriate responses at different levels are in general needed in concert to achieve desired results.

Guidance for Multiscale Assessments

Choosing the Appropriate Scales, Resolutions, and Boundaries

Several different approaches have been suggested for determining the most appropriate scale for an assessment (Wilbanks in press). One seeks the scale at which data show maximum inter-zonal variability and minimum intra-zonal variability. Another seeks the scale that minimizes statistical error between observed and modeled phenomena (Easterling et al. 1998). A third weighs increased information from finer spatial resolution against difficulties of gathering and analyzing the information (Costanza and Maxwell 1994).

The fact is, there is no single ideal scale for any instance of integrated assessment. The choice depends on the purposes of the analysis and is strongly conditioned by practical issues of data availability. The two most commonly used approaches are either to select a scale (often regional) on the basis of empirical evidence about the process involved (e.g., Kasperson et al. 1995; Schellnhuber and Wenzel 1998) or to select scales that correspond to human systems for decision-making (Cash and Moser 1998).

Despite the fact that the scale of a system is subjective (a function of the question being asked), the location of the boundaries should not be arbitrary. There are more and less useful places to locate the boundaries. The guiding principle is that a well-defined system has key feedbacks included in it and weak, slow, constant, or unidirectional interactions across the boundaries.

A practical approach to the spatial delimitation of an ecosystem is to build up a series of overlays of significant factors, mapping the location of discontinuities—for instance, in the distribution of organisms, the biophysical environment (soil types, drainage basins, shared markets), and spatial interactions (home ranges, migration patterns, fluxes of matter). A useful ecosystem boundary is one where a number of these relative discontinuities coincide. An ecosystem boundary can move over time. For instance, a marine ecosystem may be associated with an upwelling, which develops, moves, and dissipates. Similar approaches can be used to delimit human systems, such as the extent of particular patterns of land use or the political boundaries of a trade bloc. The systems addressed by the MA represent a pragmatic overlay of both ecosystem boundaries and human system boundaries.

Not all ecosystem services need to be addressed at every assessment scale. If there is a substantial mismatch between the characteristic scale of a process delivering a particular ecosystem service and the chosen scale of an assessment, it is preferable not to address that service then, leaving it to an assessment at a more appropriate scale (if such an evaluation exists).

These considerations of scale are the key reasons for performing a multiscale assessment. Comprehensive assessments need to be sensitive to multiple scales in time and space rather than focused on a single scale, and the local (or small-regional) end of the spectrum is often especially important. Processes at all relevant scales have to be included in the overall assessment, which requires appropriate methods to transfer, synthesize, and integrate information on data, variables, and processes between different scales. If results of smaller-scale studies are to be aggregated to larger scales, the prospects are brighter if they follow similar practices in the questions asked, the measurement or estimation approaches used, and the formats used for reporting results. Since time is unidirectional, temporal explicitness is typically inherent in the way observations are made and results reported. Studies must be equally explicit in the spatial dimension, although they seldom are.

The MA is designed as a collection of assessments, carried out partly independently at different scales, which are nested within one another in some cases. Effective approaches are needed for integrating top-down and bottom-up perspectives, particularly in the institutional domain, although the state of the art for such integration is not yet fully developed (Wilbanks in press).

Integration across Scales

Perhaps the greatest scale-related challenge to integrated, multiscale assessment is identifying, analyzing, and understanding linkages across scales. That they exist and are important is beyond question. The approach most often used is to analyze processes at several scales and then to examine how the findings at different levels correspond (e.g., Wu and Loucks 1995). An approach termed "strategic cyclical scaling" has been suggested in global change studies (Root and Schneider 1995). This calls for iterative cycling between upscaling and downscaling efforts, with each stage offering insights about the next as an understanding of cross-scale interactions grows.

Other suggested approaches tend to be more theoretical than practical. For instance, it is possible to think about interaction across scales as an extension of hierarchy theory (Allen and Starr 1982; O'Neill 1988). Hierarchies of scale-related processes define "constraint envelopes" within which subordinate elements of the hierarchy operate. Other possible approaches include system dynamics and dynamic spatial simulation modeling.

One relevant body of recent literature is associated with the work by the Resilience Alliance (Gunderson and Holling 2002). If systems are viewed at sufficiently long time scales, then the idea of an adaptive cycle (or configuration of states) may be valuable and can also be applied at various spatial scales.

6 Concepts of Ecosystem Value and Valuation Approaches

EXECUTIVE SUMMARY

■ Decision-making concerning ecosystems and their services can be particularly challenging because different disciplines, philosophical views, and schools of thought conceive of the value of ecosystems differently.

■ In the utilitarian (anthropocentric) concept of value, ecosystems and the services they provide have value to human societies because people derive utility from their use, either directly or indirectly (use values). People also value ecosystem services that they are not currently using (non-use values).

■ Under the utilitarian approach, numerous methodologies have been developed to try to quantify the benefits of different ecosystem services. These are particularly well developed for provisioning services, but recent work has also improved the ability to value regulating, supporting, and cultural services. The choice of valuation technique is dictated by the characteristics of each case and by data availability.

■ Non-utilitarian value proceeds from a variety of ethical, cultural, religious, and philosophical bases. These differ in the specific entities that are deemed to have value and in the interpretation of what having non-utilitarian value means. Notable among these are ecological, sociocultural, and intrinsic values. These may complement or counter-balance considerations of utilitarian value. The legal and social consequences for violating laws or regulations based on an entity's intrinsic value may be regarded as a measure of the degree of that value ascribed to them.

■ The Millennium Ecosystem Assessment plans to use valuation as a tool that enhances the ability of decision-makers to evaluate trade-offs between alternative ecosystem management regimes and courses of social actions that alter the use of ecosystems and the services they provide. This usually requires assessing the change in the mix of services provided by an ecosystem resulting from a given change in its management.

■ Most of the work involved in estimating the change in the value of ecosystem benefits concerns estimating the change in the physical flow of benefits (quantifying biophysical relations) and tracing through and quantifying a chain of causality between changes in ecosystem condition and human well-being. A common problem in valuation is that information is only available on some of the links in the chain, and often in incompatible units.

- Ecosystem values in terms of services provided are only one of the bases on which decisions on ecosystem management are and should be made. Many other factors, including notions of intrinsic value and other objectives that society might have, such as equity among different groups or generations, will also feed into the decision framework.

Introduction

The importance or "value" of ecosystems is viewed and expressed differently by different disciplines, cultural conceptions, philosophical views, and schools of thought (Goulder and Kennedy 1997). One important aim of the Millennium Ecosystem Assessment (MA) is to analyze and as much as possible quantify the importance of ecosystems to human well-being in order to make better decisions regarding the sustainable use and management of ecosystem services.

Understanding the impact of ecosystem management decisions on human well-being is an important objective. But if this information is presented solely as a list of consequences in physical terms—so much less provision of clean water, perhaps, and so much more production of crops—then the classic problem of comparing apples and oranges applies. The purpose of economic valuation is to make the disparate services provided by ecosystems comparable to each other, using a common metric. This is by no means simple, either conceptually or empirically. Society's ability to do so has increased substantially in recent years, however.

Ecosystems have value because they maintain life on Earth and the services needed to satisfy human material and nonmaterial needs. In addition, many people ascribe ecological, sociocultural, or intrinsic values to the existence of ecosystems and species. The MA recognizes these different paradigms, based on various motivations and concepts of value, along with the many valuation methods connected with them.

Ecosystems and the provisioning, regulating, cultural, and supporting services they provide have economic value to human societies because people derive utility from their actual or potential use, either directly or indirectly (known as use values). People also value ecosystem services they are not currently using (non-use values). This paradigm of value is known as the utilitarian (anthropocentric) concept and is based on the principles of humans' preference satisfaction (welfare).

Another set of values placed on ecosystems can be identified as the sociocultural perspective: people value elements in their environment based on different worldviews or conceptions of nature and society that are ethical, religious, cultural, and philosophical. These values are expressed through, for example, designation of sacred species or places, development of social rules concerning ecosystem use (for instance, "taboos"), and inspirational experiences. For many people, sociocultural identity is in part constituted by the ecosystems in which they live and on which they depend—these help determine not only how they live, but who they are. To some extent, this kind of value is captured in the concept of "cultural" ecosystem services. To the extent, however, that ecosystems are tied up with the very identity of a community, the sociocultural value of ecosystems transcends utilitarian preference satisfaction.

A different source of the value of ecosystems has been articulated by natural scientists in reference to causal relationships between parts of a system—for example, the value of a particular tree species to control erosion or the value of one species to the survival of another species or of an entire ecosystem (Farber et al. 2002). At a global scale, different ecosystems and their species play different roles in the maintenance of essential life support processes (such as energy conversion, biogeochemical cycling, and evolution). The magnitude of this ecological value is expressed through indicators such as species diversity, rarity, ecosystem integrity (health), and resilience. With increasing scarcity of space, and with limited financial resources, priorities have to be set regarding the conservation of the remaining biodiversity at all scale levels. The selection of protected areas and the determination of safe minimum standards regarding (sustainable) use of ecosystem services are based in part on these ecological values and criteria. The concept of ecological value is captured largely in the "supporting" aspect of the MA's definition of ecosystem services.

Although the various value paradigms have no common denominator and may lack any basis for comparison, some valuation approaches corresponding to them overlap and interact in various ways. Human preferences for all values can, to some extent, be measured with economic valuation methods, but ecological, sociocultural, and intrinsic value concepts have separate metrics and should be used in the decision-making process in their own right.

This chapter reviews the merits and deficiencies of these different valuation paradigms and how they complement or bound each other in assisting decisions and policy formulation for sustainable management and use of ecosystems. Ecological values are not discussed further here because they are dealt with extensively in Chapter 2.

The Utilitarian Approach and Economic Valuation Methods

The utilitarian paradigm of value is based on the fact that human beings derive utility from ecosystem services either directly or indirectly, whether currently or in the future. Two aspects of this paradigm need to be stressed. First, the use that an individual human being derives from a given ecosystem service depends on that individual's motivations, including, for example, his or her needs and personal preferences. The utilitarian approach, therefore, bases its notion of value on attempts to measure the specific usefulness that individual members of society derive from a given service, and then aggregates across all individuals, usually weighting them all equally.

Second, utility cannot be measured directly. In order to provide a common metric in which to express the benefits of the widely diverse variety of services provided by ecosystems, the utilitarian approach usually attempts to measure all services in monetary terms. This is purely a matter of convenience, however, in that it uses units that are well recognized, saves the effort of having to convert values already expressed in monetary terms into some other unit, and facilitates comparison with other activities that also contribute to well-being, such as spending on education or health. It explicitly does not mean that only services that generate monetary benefits are taken into consideration in the valuation process. On the contrary, the essence of practically all work on economic valuation of environmental and natural resources has been to find ways to measure benefits that do not enter markets and so have no directly observable monetary benefits.

Motivations for Economic Valuation

The most common reasons for undertaking a valuation of ecosystems are:

- to assess the overall contribution of ecosystems to social and economic well-being,
- to understand how and why economic actors use ecosystems as they do, and
- to assess the relative impact of alternative actions so as to help guide decision-making.

Numerous studies have assessed the contribution of ecosystems to social and economic well-being (Hartwick 1994; Asheim 1997; Costanza et al. 1997; Pimentel and Wilson 1997; Hamilton and Clemens 1999). Ecosystems form part of the total wealth of nations and contribute flow benefits, including social and cultural. But many ecosystem services are not traded, and hence their values are not captured in the conventional system of national accounts as part of total income. Moreover, in spite of the significant share of natural capital in total national wealth (World Bank 1997), the value of its depletion or appreciation is typically not accounted for.

As a result, conventional measures of wealth give incorrect indications of the state of well-being, leading to misinformed policy actions and ill-advised strategic social choices. For example, liquidation of natural assets to finance current consumption may appear to increase well-being when it does not take into account the corresponding decline in the capacity of the natural system to sustain the flow of economic, ecological, social, and cultural benefits in the future. More appropriate indicators that account for the flow and asset values of ecosystems are crucial for accurate monitoring of the implications of changes in ecosystem conditions for well-being. This is critical for the sustainable use and inter-temporal allocation of natural resources and for intergenerational equity. Valuation can help establish ecosystem values that allow correction of a country's national accounts (sometimes known as "greening") and construction of improved indicators of changes in wealth and well-being. Better valuation of the services provided by a given ecosystem does not guarantee that it will be conserved, as the costs of conservation might still be found to exceed its benefits, but it will almost certainly result in a lower loss of ecosystem services than otherwise.

Understanding why and how humans use ecosystems the way they do— for instance, why they cut natural forests, deplete soils, or pollute water surfaces—is a second reason to undertake a valuation of ecosystems. Markets guide the behavior and choices of individuals and public and private decisions. There is often a divergence, or wedge, between the market prices of goods and services as seen by individual economic agents and the social opportunity cost of using them. In particular, many services provided by ecosystems tend to be underpriced or not priced at all, leading to the inefficient and, often, unsustainable use of resources. By showing the existence and magnitude of differences between these private and social costs and benefits, valuation can help reveal policy and institutional failures (such as open access, public goods and externalities, or missing or incomplete markets), providing useful policy information on alternative intervention options for correcting them, such as creating markets or improving incentives.

FIGURE 6.1   The Total Economic Value Framework

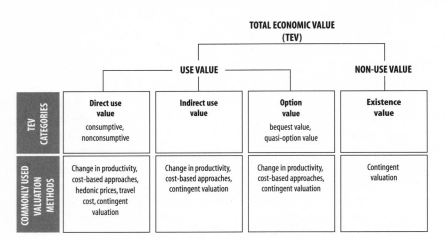

The MA plans to use valuation primarily for the third rationale for undertaking it: assessing the impacts—the gains and losses—of alternative ecosystem management regimes. This provides a tool that enhances the ability of decision-makers to evaluate trade-offs between alternative ecosystem management regimes and courses of social actions that alter the use of ecosystems and the multiple services they provide.

It must be stressed that the ecosystem values in the sense discussed in this section are only one of the bases on which decisions on ecosystem management are and should be made. Many other factors, including notions of intrinsic value, as discussed later in this chapter, and other objectives that society might have, such as equity among different groups or generations, will also feed into the decision-making framework. (See Chapter 8.)

Total Economic Value

The concept of total economic value (TEV) is a widely used framework for looking at the utilitarian value of ecosystems (Pearce and Warford 1993). (See Figure 6.1.) This framework typically disaggregates TEV into two categories: use values and non-use values.

Use value refers to the value of ecosystem services that are used by humans for consumption or production purposes. It includes tangible and intangible services of ecosystems that are either currently used directly or indirectly or that have a potential to provide future use values. The TEV separates use values as follows:

- *Direct use values.* Some ecosystem services are directly used for consumptive (when the quantity of the good available for other users is reduced) or nonconsumptive purposes (no reduction in available quantity). Harvesting of food products, timber for fuel or construction, medicinal products, and hunting of animals for consumption from natural or managed ecosystems are all examples of consumptive use. Nonconsumptive uses of ecosystem services include enjoying recreational and cultural amenities such as wildlife and bird-watching, water sports, and spiritual and social utilities that do not require a harvesting of products. This category of benefits corresponds broadly to the MA description of provisioning and cultural services.

- *Indirect use values.* A wide range of ecosystem services are used as intermediate inputs for production of final goods and services to humans such as water, soil nutrients, and pollination and biological control services for food production. Other ecosystem services contribute indirectly to the enjoyment of other final consumption amenities, such as water purification, waste assimilation, and other regulation services leading to clean air and water supplies and thus reduced health risks. This category of benefits corresponds broadly to the MA notion of regulating and supporting services.

- *Option values.* Despite the fact that people may not currently be deriving any utility from them, many ecosystem services still hold value for preserving the option to use such services in the future either by the individual (option value) or by others or heirs (bequest value). Quasi-option value is a related kind of value: it represents the value of avoiding irreversible decisions until new information reveals whether certain ecosystem services have values that are currently unknown. (Note that some analysts place option value as a subset of non-use value rather than of use value, but they do not otherwise treat it differently.) This category of benefits includes provisioning, regulating, and cultural services to the extent that they are not used now but may be used in the future.

Non-use values are also usually known as existence value (or, sometimes, conservation value or passive use value). Humans ascribe value to knowing that a resource exists, even if they never use that resource directly. This is an area of partial overlap with the non-utilitarian sources of value discussed later in this chapter. The utilitarian paradigm itself has no notion of intrinsic value. However, many people do believe that ecosystems have intrinsic value. To the extent that they do, this would be partially reflected in the existence value they place on that ecosystem, and so would be included in an assessment of its total economic value under the utilitarian approach. This kind of value is the hardest, and the most controversial, to estimate.

Economic Valuation Methods

Under the utilitarian approach, numerous methodologies have been developed to attempt to quantify the benefits of different ecosystem services (Hufschmidt et al. 1983; Braden and Kolstad 1991; Hanemann 1992; Freeman III 1993; Dixon et al. 1994). As in the case of private market goods, a common feature of all methods of economic valuation of ecosystem services is that they are founded in the theoretical axioms and principles of welfare economics. These measures of welfare change are reflected in people's willingness to pay (WTP) or willingness to accept (WTA) compensation for changes in their level of use of a particular good or bundle of goods (Hanemann 1991; Shogren and Hayes 1997). Although WTP and WTA are often treated as interchangeable, there are important conceptual and empirical differences between them. Broadly speaking, WTP is appropriate when beneficiaries do not own the resource providing the service or when service levels are being increased, while WTA is appropriate when beneficiaries own the resource providing the service or when service levels are being reduced. In practice, WTA estimates tend to be substantially higher than WTP estimates. For this reason, WTP estimates are often used, as they are more conservative.

The methods commonly used to estimate the value of various services are shown in Figure 6.1. A number of factors and conditions determine the choice of measurement method. For instance, when an ecosystem service is privately owned and traded in the market, its users have the opportunity to reveal their preferences for such a good compared with other substitutes or complementary commodities through their actual market choices, given relative prices and other economic factors. For such ecosystem services, a demand curve can be directly specified based on observed market behavior. Many ecosystem services are not privately owned or traded, however, and hence their demand curves cannot be directly observed and measured. Alternative methods have been used to derive values in these cases. Different users and authors often classify the various methods of measuring ecosystem services values differently, but the grouping and naming systems converge to a broad classification that basically depends on whether the measures are based on observed or hypothetical behavior.

BOX 6.1   Valuation of Economic Services Through Observed Behavior

---

■ *Direct observed behavior methods*. These methods derive estimates of value from the observed behavior of producers and consumers. They often use market prices and are most often applicable in cases where the ecosystem services are privately owned and traded in functioning markets. This approach is most frequently applicable to consumptive use, where goods are extracted from ecosystems and traded on markets.

■ *Indirect observed behavior methods*. This category also uses actual observed behavior data but not on the ecosystem service in question. In the absence of actual market behavior regarding that particular service, these methods use observations on actual behavior in a surrogate market, which is hypothesized to have a direct relationship with the ecosystem service value. Examples in this category include hedonic pricing methods (which use statistical techniques to break down the price paid for a service into the implicit prices for each of its attributes, including environmental attributes such as access to recreation or clean air) and travel cost methods (which use observed costs to travel to a destination to derive demand functions for that destination). This group also includes cost-based methods (such as replacement cost methods, which value services at the cost of replacing, for example, a water purification service provided by an ecosystem with a new water treatment plant) that do not exactly reflect welfare (benefit-based) measures of value. (They sometimes underestimate and sometimes overestimate value.)

The standard valuation approach that uses actual observed behavior data is further divided into direct and indirect observed behavior methods. (See Box 6.1.) When they can be applied, these are generally considered preferable to measures based on hypothetical behavior.

The second valuation approach uses measures of economic value based on hypothetical behavior. In this category of methods, people's responses to direct questions describing hypothetical markets or situations are used to infer value. This group can also be divided into direct hypothetical (such as contingent valuation, in which respondents are asked directly how much they would be willing to pay for specified benefits) and indirect hypothetical measures of WTP or WTA (contingent ranking or conjoint valuation, which ask respondents to rank different bundles of goods).

A final category of approach is known as benefits transfer. This is not a methodology per se but rather the use of estimates obtained (by whatever method) in one context to estimate values in a different context. For example, an estimate of the benefit obtained by tourists viewing wildlife in one park might be used to estimate the benefit obtained from viewing wildlife in a different park. Benefits transfer has been the subject of considerable controversy in the economics literature, as it has often been used inappropriately. A consensus seems to be emerging that benefits transfer can provide valid and reliable estimates under certain conditions. These include that the commodity or service being valued is identical at the site where the estimates were made and the site where they are applied and that the populations affected have identical characteristics. Of course, the original estimates being transferred must themselves be reliable for any attempt at transfer to be meaningful.

Each of these approaches has seen broad use in recent years, and an extensive literature exists on their application. These techniques can and have been applied to a wide range of issues, including the valuation of cultural benefits (Pagiola 1996; Navrud and Ready 2002). In general, more direct measures are preferred to indirect ones. However, the choice of valuation technique in any given instance will be dictated by the characteristics of the case and by data availability.

Several techniques have been specifically developed to cater to the characteristics of particular problems. The travel cost method, for example, was developed to measure the utility derived by visitors to sites such as protected areas. The change in productivity approach, on the other hand, is quite broadly applicable to a wide range of issues. Contingent valuation is potentially applicable to any issue, simply by phrasing the questions appropriately, and as such has become widely used—probably excessively so, as it is easy to misapply and, being based on hypothetical behavior, is inherently less reliable. Data availability is a frequent constraint and often restricts the choice of approach. Hedonic price techniques, for instance, require vast amounts of data, thus limiting their applicability.

Putting Economic Valuation into Practice

Whichever method is used for valuing a service, the analysis must begin by framing appropriately the question to be answered. In most policy-relevant cases, the concern is over changes in the level and mix of services provided by an ecosystem. At any given time, an ecosystem provides a specific "flow" of services, depending on the type of ecosystem, its condition (the "stock" of the resource), how it is managed, and its socioeconomic context. A change in management (whether negative, such as deforestation, or positive, such as an improvement in logging practices) will change the condition of the ecosystem and hence the flow of benefits it is capable of generating. It is rare for all ecosystem services to be lost entirely; a forested watershed that is logged and converted to agriculture, for example, may still provide a mix of provisioning, regulating, supporting, and cultural services, even though both the mix and the magnitude of specific services will have changed. Consequently, an assessment of the change in the value of services resulting from a given change in ecosystem management typically is most relevant to decision-makers and policy-makers. Where the change does involve the complete elimination of eco-system services, such as the conversion of an ecosystem through urban expansion or road-building, then the change in value would equal the total economic value of the services provided by the ecosystem. (Measure-ments of total economic value of the services from a particular ecosystem can also be useful to policy-makers as an economic indicator, just as mea-sures of gross national product or genuine savings provide policy-relevant information on the state of the economy.)

An assessment of the change in value of ecosystem services can be achieved either by explicitly estimating the change in value or by sepa-rately estimating the value of ecosystem services under the current and the alternative management regime and then comparing them. If the loss of a given service is irreversible, then the loss of the option value of that service will also be included. (An important caveat here is that the appro-priate comparison is between the ecosystem with and without the man-agement change; this is not the same as a comparison of the ecosystem before and after the management change, as many other factors will usu-ally also have changed.) The typical question being asked, then, is whether the total value of the mix of services provided by an ecosystem managed in one way is greater or smaller then the total value of the mix provided by that ecosystem managed in another way.

The actual change in the value of the benefits can be expressed either as a change in the value of the annual flow of benefits, if these flows are relatively constant, or as a change in the present value of all future flows. The latter is equivalent to the change in the capital value of the ecosys-tem and is particularly useful when future flows are likely to vary substan-tially over time. (It is important to bear in mind that the capital value of the ecosystem is not separate and additional to the value of the flows of benefits it generates; rather, the two are intimately linked in that the capi-tal value is the present value of all future flows of benefits.)

Estimating the change in the value of the flow of benefits provided by an ecosystem begins by estimating the change in the physical flow of ben-efits. This is illustrated in Figure 6.2 for a hypothetical case of deforesta-tion that affects the water services provided by a forest ecosystem.

FIGURE 6.2   Valuing the Impact of Ecosystem Change

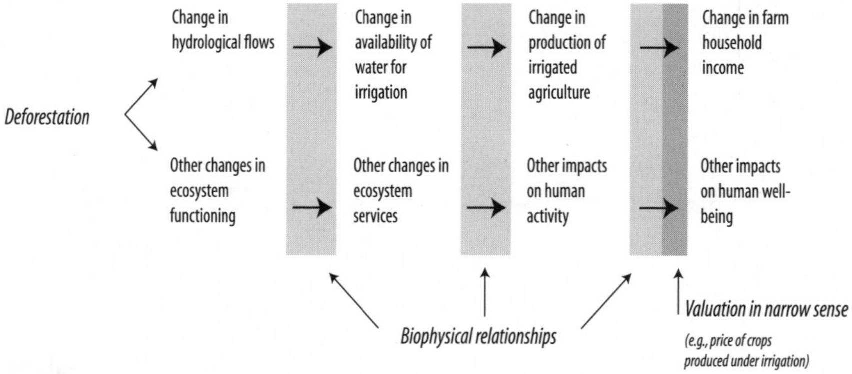

Source: Adapted from Pagiola et al. in press.

The bulk of the work involved in the exercise actually concerns quantifying the biophysical relationships. In many cases, this requires tracing through and quantifying a chain of causality. Thus, valuing the change in production of irrigated agriculture resulting from deforestation requires estimating the impact of deforestation on hydrological flows, determining how changes in water flows affect the availability of water to irrigation, and then estimating how changes in water availability affects agricultural production. Only at the end of this chain does valuation in the strict sense occur—when putting a value on the change in agricultural production, which in this instance is likely to be quite simple, as it is based on observed prices of crops and agricultural inputs. The change in value resulting from deforestation then requires summing across all the impacts.

Clearly, following through a chain like this requires close collaboration between experts in different disciplines—in this example, between foresters, hydrologists, water engineers, and agronomists as well as economists. It is a common problem in valuation that information is only available on some of the links in the chain, and often in incompatible units. The MA can make a major contribution by helping the various disciplines involved to become more aware of what is needed to ensure that their work can be combined with that of others to allow a full analysis of such problems.

In bringing the various strands of the analysis together, there are many possible pitfalls to be wary of. Inevitably, some types of value will prove impossible to estimate using any of the available techniques, either because of lack of data or because of the difficulty of extracting the desired information from them. To this extent, estimates of value will be underestimates. Conversely, there is an opposite danger that benefits (even if accurately measured) might be double-counted.

As needed, the analysis can be carried out either from the perspective of society as a whole ("social" analysis) or from that of individual groups within society ("private" analysis). Focusing on a particular group usually requires focusing on a subset of the benefits provided by an ecosystem, as that group may receive some benefits but not others. (Groups located within an ecosystem, for example, typically receive most of the direct use benefits but few of the indirect use benefits, whereas the opposite applies to downstream users.) It will often also require using estimates of value specific to that group; the value of additional water, for example, will be different depending on whether it is used for human consumption or for irrigation. The analysis can thus allow distributional impacts and equity considerations to be taken into account, as well as overall welfare impacts on society as a whole. This type of disaggregation is also useful in understanding the incentives that particular groups face in making their ecosystem management decisions. Many ecosystems are mismanaged, from a social perspective, precisely because most groups that make decisions about management perceive only a subset of the benefits the ecosystem provides.

Similarly, estimating the impact of changes in management on future flows of benefits allows for intergenerational considerations to be taken into account. Here, too, the bulk of the work involved concerns predicting the change in future physical flows; the actual valuation in the narrow sense forms only a small part of the work. Predicting the value that future generations will place on a given service is obviously difficult. Technical, cultural, or other changes could result in the value currently placed on a service either increasing or decreasing. Often, the best that can be done is to simply assume that current values will remain unchanged. If trends suggest that a particular change in values will occur, that can be easily included in the analysis. Such predictions are notoriously unreliable, however.

Non-utilitarian Value

From the perspective of many ethical, religious, and cultural points of view, ecosystems are valued even if they do not contribute directly to human well-being. Some ecosystems may be vital to a people's identity as a distinct society or culture. Thus preserving the health of such ecosystems may be a necessary condition for measuring changes in the collective welfare of those societies and cultures. Further, to the extent that a society's or a culture's ecocentric philosophical and ethical views recognize the intrinsic value of nonhuman species and ecosystems, sociocultural value also reaches beyond human welfare considerations.

Sociocultural Values

For many people, ecosystems are closely associated with deeply held historical, national, ethical, religious, and spiritual values. A particular mountain, forest, or watershed may, for example, have been the site of an important event in their past, the home or shrine of a deity, the place of a moment of moral transformation, or the embodiment of national ideals. These are some of the kind of values that the MA recognizes as the cultural services of ecosystems. And to some extent they are captured by utilitarian methods of valuation. But to the extent that some ecosystems are essential to a peoples' very identity, they are not fully captured by such techniques.

These values fall between the utilitarian and intrinsic value paradigms. They might be elicited by using, for example, techniques of participatory assessment (Campell and Luckert 2002) or group valuation (Jacobs 1997; Wilson and Howarth 2002). This evolving set of techniques is founded on the assumption that the valuation of ecological goods and services should result from a process of open public deliberation, not from the aggregation of separately measured individual preferences. Using this approach, small groups of citizens are brought together in a moderated forum to deliberate about the economic value of ecosystem goods or services (Wilson and Howarth 2002). The end result is a deliberative or "group" contingent valuation (CV) process (Jacobs 1997; Sagoff 1998). With a group CV, the explicit goal is to derive an economic value for the ecological good or service in question. The valuation exercise is conducted in a manner very similar to a conventional CV survey—using hypothetical scenarios and payment vehicles—with the key difference being that value elicitation is not done through private questioning but through group discussion and consensus building.

The Intrinsic Value Paradigm

Although the notion that nature has intrinsic value is a familiar one in many religions and cultures, it is unfamiliar in the context of modern rational choice theory and economic valuation. Yet analysts do have a well-established and familiar metric for assessing the intrinsic value of human beings and their various aspects. This valuation method and its metric may then be extended to some nonhuman natural entities, including eco-systems.

The notion that ecosystems have intrinsic value is based on a variety of points of view. Intrinsic value is a basic and general concept that is founded upon many and diverse cultural and religious worldviews. Among these are indigenous North and South American, African, and Australian cultural worldviews, as well as the major religious traditions of Europe, the Middle East, and Asia.

In the Judeo-Christian-Islamic tradition of religions, human beings are alleged to be created in the image of God. On that basis, humans are attributed intrinsic value. The Bible also represents God as having created plant and animal species, and declares the things thus created to be "good." Some commentators have argued that in doing so, God attributes intrinsic value to them, and thus that plant and animal species and the other aspects of nature that God also declared to be good have intrinsic value by an act of divine fiat (Barr 1972; Zaidi 1981; Ehrenfeld and Bently 1985).

In some American Indian cultural worldviews, animals, plants, and other aspects of nature are conceived as relatives, born of one universal Mother Earth and Father Sky (Hughes 1983). Thus they have the same value as human relatives: intrinsic value—if not in name, then at least in pragmatic effect. You may not sell your mother at any price; even perform-ing a hypothetical economic valuation of your mother is questionable. And so, some American Indian elders have argued, neither should hu-mans sell Mother Earth—that is, their tribal lands—or even compromise the intrinsic value of Earth by carrying out an economic valuation of tribal lands (Gill 1987).

Examples of other religious worldviews supporting the concept of in-trinsic value in nature abound. Basic to Hindu religious belief is the essen-tial oneness of all being, Brahman, which lies at the core of all natural things. The presence of Brahman in all natural things is the Hindu basis of intrinsic value (Deutch 1970). Closely related to this idea is the moral imperative of *ahimsa,* non-injury, extended to all living beings. The con-cept of *ahimsa* is also central to the Jain environmental ethic (Chapple 1986). Buddhism incorporates *ahimsa* as a central moral imperative as well (Chapple 1986). Also central to Buddhism is the overcoming of suffering by the cessation of desire. Absent desire, the natural world ceases to be referenced to a person as a pool of resources existing to satisfy desires or preferences (Kalupahana 1985). The enlightened Buddhist is thus able to appreciate the intrinsic value of nature.

Taoism, a major philosophical and religious tradition of China, posits the *Tao* or Way of nature as a norm of human action (Tu 1985). Taoism regards human economies as a subset of the economy of nature. In the Japanese Shinto religious tradition, the *kami* (gods), are closely associated with various aspects of nature (Odin 1991). As the *kami* have a greater-than-human dignity, the aspects of nature with which they are associated are also thought to have intrinsic value. In the Dreamtime narratives of the peoples indigenous to Australia, various features of the landscape are the places where the totemic Ancestors performed "terraforming" deeds (Stanner 1979). Such places are sacred and, in effect, have intrinsic value.

These are but a few of the bases for intrinsic value in non-western religious and cultural worldviews (for a comprehensive summary, see Callicott 1994). It is important for decision-makers to assess empirically the actual ecosystem-oriented values—intrinsic, sociocultural, and ecological, as well as utilitarian—of those affected by ecosystem-oriented policy and decisions.

The two main traditions of modern secular ethics in western culture are utilitarianism and Kantianism. In classical utilitarianism, aggregate "happiness," understood as a greater balance of pleasure over pain, was the putative goal of social policy. Contemporary economics is derived from utilitarianism and posits "preference satisfaction" as the goal of rational choice (Sen 1987). If aggregate preference satisfaction is, correspondingly, the goal of social policy, this may sometimes be maximized at the cost of overriding the interests of a comparatively few individuals (Rawls 1971). The potential injustices of unbridled utilitarianism are checked by the assertion of individual rights—most basically to life, liberty, and property.

Economic valuation of ecosystem services has been variously criticized by different commentators (e.g., Bromley 1990; Costanza 2000; Heal 2000a; Heal 2000b; Ludwig 2000; Pritchard et al. 2000). Further, reducing all values to preferences has been contested (Sagoff 1988). A person may prefer chocolate to vanilla ice cream, but some find it demeaning to the intrinsic value of human life and human liberty to say that as a society humans collectively prefer not to stage gladiator shows or own slaves or that, as an individual, a person merely prefers honesty over perfidy or justice over treachery.

The counter-utilitarian idea that there is a difference between preferences and values and that considerations of individual rights tempers calculations of aggregate utility was most clearly and powerfully expressed by Kant, who wrote, "Everything has either a *price* or a *dignity*. Whatever has a price can be replaced by something else as its equivalent; on the other hand, whatever is above all price, and therefore admits of no equivalent, has a dignity. But that which constitutes the condition under which alone something can be an end in itself does not have mere relative worth, i.e., a price, but an intrinsic worth, i.e., a dignity" (Kant 1959 [1785]:53, italics in original).

Because human rights, based on the dignity and intrinsic value of human beings, has traditionally been used to check the excesses and potential injustices of calculations of aggregate utility, many non-anthropocentric ethical theorists have largely adopted the intrinsic value paradigm. They first extended it to cover various nonhuman animals (Regan 1983). Some have attempted to push this line of argument further, to argue that all organisms have interests, goods of their own, natural goals, developments, and fulfillments and so should be accorded intrinsic value (Taylor 1986). Based on the seminal work of Aldo Leopold (1949), others have argued that transorganismic levels of biological organization (species, biotic communities, ecosystems) also have intrinsic value (Callicott 1989; Rolston III 1994). On whatever basis, intrinsic value has been attributed to various aspects of nature (genes, organisms, populations, species, evolutionarily significant units, biotic communities, ecosystems) and to nature as a whole (the biosphere).

The basis on which intrinsic value is attributed to various entities may limit which ones can have intrinsic value. For example, if being rational is the property required for something to have intrinsic value, then only rational beings (effectively, only human beings) are recognized to be intrinsically valuable. Non-anthropocentric theorists who have posited the criterion of "having interests" for ascribing intrinsic value thus limit it to individual organisms. In traditional Judeo-Christian thinking, those who thought that intrinsic value should be based on the property of being created in the image of God also effectively limit intrinsic value to human beings. In the Dreamtime worldview of the peoples indigenous to Australia, although landscape-level features have intrinsic value, individual plants and animals usually do not (except those associated with a person's own totem). Aldo Leopold (1949) thought that the things deserving of human "love and respect" had intrinsic value. Theoretically someone can love and respect anything at all, but Leopold argued that among other things, "biotic communities" commended themselves to human capacity for love and respect.

The Interactions of Political and Market Metrics

Parallel to using the market or its surrogates to measure economic value, in democratic societies the modern social domain for the ascription of intrinsic value is the parliament or legislature (Sagoff 1988). In other so-cieties a sovereign power ascribes intrinsic value, although this may less accurately reflect the actual values of citizens than parliamentary or legis-lative acts and regulations do. The metric for assessing intrinsic value is the severity of the social and legal consequences for violating laws prohib-iting a market in or otherwise compromising that which is recognized to be intrinsically valuable. In western societies long influenced by the Judeo-Christian worldview and Kantian moral philosophy, the highest intrinsic value is attributed to human life. Thus the severest of consequences are prescribed for murdering human beings.

Each kind of value—utilitarian, ecological, sociocultural, and intrin-sic—is played out on a common and not always level playing field. Thus the various kinds of value intersect and interact in various ways. One com-mon effect of socially recognizing and legally institutionalizing something's intrinsic value is to take it off the open market, to insist that it has a dignity and therefore should have no price. The clearest and most obvious example is human beings themselves. In most modern societies, there is no legal market in human beings; there is no open slave market. With the advent of human organ transplants, some societies have decided that there should be no legal market in human organs either; these are, by implica-tion, thus accorded intrinsic value.

A black market often emerges in entities that are sufficiently well rec-ognized as having a dignity to register a signal in the political intrinsic value metric. Depending on the strength of that signal—for instance, the social and legal consequences of pricing and trafficking in that entity—the supply of such entities declines and the price rises. So one effect of the political intrinsic value metric on the market metric is analogous to the effect of an excise tax or tariff.

Some things may arguably have both a dignity and a price—human labor, for example. Society may protect the recognized intrinsic value of things that also have utility by assuring, among other things, that their price is right. This may be the ethical rationale for minimum-wage laws, legally mandated health insurance, and retirement benefits in societies that have provided such protections by law. Society may also constrain the use of human labor with regulations designed to protect workers' health and safety.

Laws and regulations recognizing the intrinsic value of such things as endangered species, biodiversity more generally, and ecosystems such as wetlands have created a regulatory environment to which market forces are beginning to respond. A legal market in conservation "credits" is emerging. The red-cockaded woodpecker, for example, is a "listed" species protected by the U.S. Endangered Species Act (ESA), administered by the U.S. Fish and Wildlife Service (FWS). An agreement between International Paper (IP) and the FWS permits the company to consolidate at one location the breeding pairs of red-cockaded woodpeckers on its properties in several southeastern states and intensively manage that location as habitat for the endangered species. The agreement permits IP to harvest timber on the vacated sites and to sell credits to other owners of red-cockaded woodpecker habitat as the species recovers and the number of breeding pairs increases beyond a specified threshold (U.S. Fish and Wildlife Service 1999). Similarly, a company wishing to convert a wetland to a shopping mall faces regulatory constraints prohibiting wetland destruction. It can comply with those constraints by purchasing a credit from a distant landowner whose property contains a comparable wetland that will be protected (Fernandez 1999). This provides a market incentive to wetlands owners to conserve them.

Another effect of the political intrinsic value metric on the market metric is to shift the burden of proof away from those who would protect something with socially recognized and legally sanctioned intrinsic value and toward those who would commercially exploit it. The debate about human embryonic stem-cell research in the United States is a case in point. As aspects of human being, human embryonic stem cells are alleged to have a dignity and therefore should not be commercially exploited by the pharmaceutical industry, some have argued (with ambiguous political success). To overcome this argument, the pharmaceutical industry and its scientific allies must successfully counterargue that the aggregate utility of human embryonic stem-cell research is so great as to warrant overriding the putative dignity of this aspect of human being (Orkin and Morrison 2002).

Just because something has publicly recognized intrinsic value does not mean that its value is absolute or inviolable. Even human beings can be "converted" in deference to other values. Soldiers, for example, are often placed in harm's way to advance a country's perceived national interests or even aggregate economic welfare. In such cases, the intrinsic value of human beings seems sacrificed in favor of other values. But when intrinsic values are in zero-sum conflict with utilitarian values, the burden of proof rests with those advocating the latter.

Perhaps the most interesting and relevant case in point of legislative ascription of intrinsic value to some aspect of nature—and of the meeting of utilitarian and intrinsic value metrics—is the U.S. Endangered Species

Act enacted in 1973. In giving absolute legal protection to listed endangered species, the ESA, in effect, gave them a dignity comparable in strength to the dignity accorded individual human life. As noted, even the dignity of human life can be legally overridden, but the burden of proof falls on those who would do so. The ESA was amended in 1978 to create a Cabinet-level Endangered Species Committee empowered to decide whether opportunity cost (measured on the market metric) of protecting a listed species was high enough to warrant overriding its dignity (measured on the political metric).

This interaction between the political metric of intrinsic value and the market metric (and its surrogates) of utilitarian value has an analog in economic valuation called the safe minimum standard (SMS). Approaching the task of economically valuing ecosystem services by means of the SMS is practically equivalent to socially recognizing their intrinsic value and protecting them by law. Whereas benefit-cost analysis approaches each case and builds up a body of evidence about the benefits and costs of preservation, the SMS approach starts with a presumption that the maintenance of the healthy functioning of any ecosystem is a positive good (lumping together economic, ecological, sociocultural, and intrinsic values). The empirical economic question is, How high are the opportunity costs of satisfying the SMS? The SMS decision rule is to maintain the ecosystem unless the opportunity costs of doing so are intolerably high. The burden of proof is thus assigned to the case against maintaining the SMS (Randall 1998).

The quantitative threshold to which the opportunity costs must rise to warrant violating the SMS is left as an open empirical question. In practice, such thresholds are set by the political metric. The economic threshold for violating the SMS for ecosystem health will depend in part on how successful its advocates are in persuading voters that ecosystems have a dignity—not necessarily instead of, but as well as a price—and should be protected unless the opportunity costs of doing so are intolerably high. The question of how high is high enough will be indicated in part by the strength of laws and regulations enacted to protect ecosystems. In this case, however, the intrinsic value (assessed on the political metric) is augmented by the considerable utilitarian value of ecosystem services; their psycho-spiritual utilitarian values; their option, bequest, and existence utilitarian values; and their ecological and sociocultural values.

Conclusion

Human societies face important choices in how they manage ecosystems, affecting their conditions and the services they provide and thus ultimately human well-being. How decisions are made will depend on the systems of value endorsed in each society, the conceptual tools and methods at their disposal, and the information available. Making the appropriate choices requires, among other things, reliable information on actual conditions and trends of ecosystems and on the economic, political, social, and cultural consequences of alternative courses of action.

The MA will provide decision-makers with relevant information to aid them in making appropriate ecosystem management decisions. The impact that these decisions will have on human well-being is of particular interest. In some cases, these impacts can be assessed with indicators, such as the impact on human health. When there are multiple impacts and well-being is affected in many different ways, however, such unidimensional indicators will not be sufficient. In these cases, economic valuation will provide an important tool, as it will allow for different impacts to be compared and aggregated.

Of course, the importance of ecosystems goes beyond their role for human well-being. Non-utilitarian sources of value must also be taken into consideration in order to make appropriate management decisions.

7 Analytical Approaches

Executive Summary

- The overall analytical approach to be used to achieve the goals of the Millennium Ecosystem Assessment (MA) has nine major tasks: identifying and categorizing ecosystems and their services; identifying links between human societies and ecosystem services; identifying the direct and indirect drivers of change; selecting indicators of ecosystem conditions, services, human well-being, and drivers; assessing historical trends and the current state of ecosystems, services, and drivers; evaluating the impact of a change in services on human well-being; developing scenarios of ecosystems, services, and drivers; evaluating response options to deal with ecosystem changes and human well-being; and analyzing and communicating the uncertainty of assessment findings.

- The MA will rely on five major categories of data and indicators: core data sets (shared among all MA Working Groups), data and indicators for assessment reports (closely targeted to individual analyses), indicators for summary and synthesis reports (a smaller set of clear, policy-relevant indicators), new data sets (developed during the MA process for continued use), and metadata (data documenting all of these data sets).

- Although new synoptic data sets (for example, from remote sensing) enable more comprehensive global assessments, they nevertheless have deficiencies that need to be addressed. These include incomplete and inconsistent spatial and temporal data coverage, contradictory definitions of types of data, and the mismatch of ecological, geographic, and political boundaries. Some of these deficiencies will be addressed when the MA acts to assure the quality of data used in the assessment. Various steps could be taken for data quality assurance, such as setting up a data archive, sponsoring the development of MA data sets, or using data already described in the scientific literature.

- Models will play an integrative role in the MA and will complement data collection and analysis. Modeling will be used to analyze interactions among processes, fill data gaps, identify regions for priority data collection, and synthesize existing observations into appropriate indicators.

- The MA will develop four or five scenarios of medium- to long-term changes in ecosystems, services, and drivers. The scenarios will have an explicitly ecological perspective and will explore such themes as ecological surprises and cross-scale ecological feedbacks. They will build on the social and economic information contained in existing global scenarios.

■ Scientists must make every effort to estimate the certainty of important find-ings. They must then distinguish and communicate which findings are robust, which are partially understood, and which are uncertain or even speculative. As a rule, uncertainties from all aspects of an assessment should be reported in a consistent and transparent way.

Introduction

The analytical approach used to achieve the goals of the Millennium Eco-system Assessment (MA) must be suitable to the many disciplines in-volved in the MA and address the MA conceptual framework, synthesiz-ing the state of knowledge concerning the impact of ecosystem changes on human well-being. The management, analysis, and interpretation of information are key issues because of their relevance to maintaining high scientific standards in the assessment and because they can facilitate the accessibility and usefulness of MA results. Moreover, the effective man-agement of information is a vital requirement for providing a scientific record of a comprehensive global assessment of the world's ecosystems.

There are nine major tasks in the analytical approach of the MA. (See Figure 7.1.) Note that few arrows are shown in Figure 7.1 to emphasize

FIGURE 7.1   The Analytical Approach of the Millennium Ecosystem Assessment and Its Main Tasks

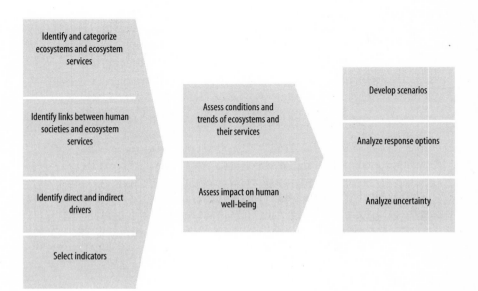

that many of the activities will be carried out simultaneously rather than in sequence, although at some junctures information will feed in from one task to another.

- *Identify and categorize ecosystems and their attendant services.* To facilitate the assessment of complex ecosystems, the MA will classify them into a limited number of categories as a basis for assessing the services they provide. Ecosystem services are identified and grouped into functional categories: provisioning, regulating, cultural, and supporting. (See Chapter 2.)

- *Identify links between services and human societies.* Here the links are described between human societies and the particular ecosystem services that they use or benefit from. This includes defining the components of human well-being that are affected by the services (such as health, livelihood, culture, and equity), as well as the human activities that in turn affect ecosystems and the supply of services (such as population growth, consumption, and governance). (See Chapter 3.)

- *Identify indirect and direct drivers.* In this task a list of indirect and direct drivers of the state of ecosystems and their services is drafted. Indirect and direct drivers affect not only ecosystems and their services but also each other. For example, demographic changes (an indirect driver) can affect ecosystems though land use change (a direct driver) but also can influence other indirect drivers such as social values and institutions. (See Chapter 4.)

- *Select indicators of ecosystem conditions, services, human well-being, and drivers.* A set of indicators is selected to assess the state of ecosystems, ecosystem services, human well-being, and drivers. As an example, if the ecosystem service is food provision, then a potential indicator for the ecosystem state would be area under cultivation; for the service, quantity of food produced; for human well-being, rates of malnutrition; and for drivers, population growth. Next, these indicators are quantified or otherwise evaluated for use in the other analytical tasks. (See Chapters 2, 3, and 4.)

- *Assess historical trends and the current state of ecosystems and their services and drivers.* The current state of ecosystems and their services is assessed by assembling and analyzing data on the indicators selected. The details of how these data will be analyzed have not been completely worked out, but some considerations are discussed in Chapter 2. Since ecosystems are dynamic, an important issue to be addressed is the meaning of "current conditions." In some cases this will refer to the most recent data collected, but for most ecosystems it must take into account year-to-year and perhaps inter-decadal variability. (For example, it is not useful to refer to the availability of fresh water for a particular year because of its strong year-to-year variability.)

- *Evaluate impact on human well-being.* This is among the most challenging tasks in the MA, since it involves the translation of information largely from the natural sciences (such as the state of fresh water, soil, and forests) into variables of concern to society (health, livelihoods, wealth, and security, for instance). One challenge is that a given service can affect several components of human well-being. Another challenge lies in sorting out the many possible trade-offs among services. Finally, the distribution of service benefits among societal groups will need careful consideration.

- *Develop scenarios.* The MA is concerned not only with the historical, present, and short-term future trends of ecosystems, but also with future trends over the medium and longer term. This information is needed to anticipate critical changes in ecosystems and to develop response strategies. The aim of this task is to identify a set of plausible futures or "scenarios" for ecosystems, services, and drivers.

- *Evaluate possible responses.* In this task the many possible "response options" are identified for preventing the deterioration of ecosystem services or recovering lost services. This includes evaluating the success of past response options and developing guiding principles for designing needed policies. Consistency is needed between the response strategies identified here and those used in the scenarios. (See Chapter 8.)

- *Analyze and communicate uncertainty.* Since the MA is concerned with a new and rapidly changing body of knowledge, it is clear that many of the findings will be uncertain. Assessing and communicating the level of certainty in a clear and consistent manner is therefore a central task of the MA.

These nine tasks and Figure 7.1 do not pertain to any particular spatial or temporal scale. Nevertheless, assessments carried out on the sub-global scale might require some refinement of the tasks. For example, in a sub-global assessment the selection of ecosystem categories must take into account the unique conditions of a region, such as its existing biogeographic zones. Another example is the selection of indirect and direct drivers, which should reflect the relevant temporal and spatial scales of the assessment, while also taking into account possible external global drivers. As a general rule, the nine MA tasks described should be adjusted to the particular needs of each sub-global assessment.

At the global scale, the MA has distributed these tasks among three Working Groups. The Condition and Trends Working Group is concerned with the first six tasks, the Scenarios Working Group builds on these to focus on the seventh task, and the Responses Working Group builds on all the earlier tasks to focus on the eighth task. All three Working Groups are centrally concerned with the analysis and presentation of uncertainty and with incorporating uncertainty into decision-making.

The MA Working Groups also focus on distinct time intervals. The Condition and Trends Working Group will assess current conditions and historic trends, typically over the last 40 years. This group will also consider issues of sustainability, presenting short-term projections (typically over the next 10 years) of changes in ecosystems, ecosystem services, and associated human well-being. The Scenarios Working Group will consider plausible futures over the next 25, 50, and 100 years. The Responses Working Group will assess the success of past and current responses and will use these assessments to evaluate available future responses.

The conceptual issues surrounding these nine tasks are discussed further in previous chapters, and the specific methodologies involved in accomplishing them will be better described and applied in the Working Group reports. The remainder of this chapter describes several of the major cross-cutting issues in the MA analytical approach:

- data,
- units of analysis and reporting,
- modeling,
- scenarios, and
- scale and uncertainty.

Data

A global assessment of world ecosystems and their services obviously requires an enormous amount of data. These needs have been summed up into five broad categories in the MA sub-group report, *Core Data Sets and Indicators*:

- *Core data sets.* Core data sets are those with wide potential application in the MA. They could cover, for example, land use, land cover, freshwater resources, marine resources, population, and infrastructure. Establishing common core data sets for use by all Working Groups and scientists within the MA will maximize consistency among analyses. In general, the MA will ensure timely access for all participants to such core data sets via an online data archive. Core data sets could be data already available or data developed specifically for an assessment.

- *Data and indicators for assessment reports.* Each chapter of the MA will necessarily make extensive use of published data and indicators. In addition, it is likely that some chapters will develop new indicators to meet their particular needs, recalculate existing indicators based on the agreed-upon core data sets (for example, recalculate a measure of fish production based on an updated marine ecosystem classification), or extend indicators developed for specific regions to the global scale.

- *Indicators for summary and synthesis reports.* Compared with the many indicators used in the full MA reports, only a small number can be included in the Summaries for Decision-makers or the Synthesis Reports. These key indicators (perhaps 10–15) either will be selected from the larger number or will be compound indicators incorporating several others. An enormous weight will fall on these indicators in communicating the core MA findings to decision-makers. They must generally be highly relevant to policy-makers, easily understood, and effectively convey the bottom-line findings concerning the consequences of ecosystem change on human well-being. Given the pivotal role these indicators will play with respect to the perception and impact of the MA beyond the scientific community, they will be explicitly identified and targeted for development.

- *New data sets.* The existence of the MA will probably stimulate the production of new data sets that may be less useful to the MA itself (because of their timing, perhaps, or their resolution) yet would be valuable for other institutions. These data sets could be helpful, for example, in building the capacity of institutions to undertake their own regional, national, or sub-national integrated assessments of ecosystems and their services. For instance, the United States has promised to provide the MA with complete global terrestrial cover from Landsat 7 for the year 2000. Although it is unlikely these data will be fully available in time for the MA, they will ensure that geo-referenced Landsat data will be available at low cost to any country or institution interested in undertaking a more fine-grained analysis of land cover change.

- *Metadata.* For both scientific and technical purposes, it is important to document the data used in the MA (so-called metadata) and to make the documentation widely available. This need arises in part from the scientific responsibility to make the work of the assessment transparent, traceable, and reproducible. But there are also reasons of data management, given the breadth and diversity of data used in global assessments (for instance, gathering sufficient information to define the origin of the data and assess its reliability). It is becoming more common that software behind, for example, geographic information systems and Web browsers uses standardized metadata descriptions to organize and search for information. Standards for metadata should include the number and format of data description fields, original data grain and extent, and the selection of appropriate searchable keywords.

To facilitate access to both core data sets and metadata, the MA will establish a data archive. The archive will receive computer support over several years and have the appropriate technical characteristics for conveniently storing and transferring large data sets.

Challenges in Using Data

There has been a recent proliferation of data sets of differing geographic extent relevant to the work of the MA. They describe the location, extent, and condition of ecosystems, the provision of ecosystem services, and, less frequently, the relationships among drivers and ecosystem services or among ecosystem services and human well-being. Some of these are based on remote sensing and other relatively recent technologies, while others are from new field programs. These data sets will allow the MA to conduct more rigorous, inclusive, and globally consistent assessments than would have been possible perhaps 10–15 years ago.

Nevertheless, the MA faces several difficult issues in using these data effectively. First, the data are incomplete in coverage and are often collected by many different researchers who sometimes use incompatible methods. Second, data often have inconsistent spatial scales and time periods, use distinctive definitions and characterization approaches, and are rarely adequately documented, particularly in terms of describing the accuracy and reliability of data sets and models. Third, the reality is that widely accepted data sets for many important aspects of the world's ecosystems are simply not available. For example, land cover derived from different global data sources (different remote sensing instruments and ground-truthing techniques) often provides conflicting information, none of them match national land use statistics, and time series data of global land cover have never been produced.

Perhaps the greatest challenge is that the MA aims to be a global and integrated assessment, yet the available and relevant data continue to be of uneven quality in terms of geographic and temporal extent as well as resolution, taxonomy, and economic sector. (For example, crop data are of generally better quality than fisheries or livestock data, which in turn are likely to be better than fuelwood or biodiversity statistics.) Unfortunately, from a geographical perspective, the completeness and reliability of data are often inversely related to the rate of ecosystem change and to levels of human welfare.

Data on subjects as varied as species diversity patterns, deforestation rates, invasive plant distributions, human demographic trends, and economic indicators are often more accessible, of greater reliability, and of higher spatial and temporal resolution in richer countries. For example, reliable estimates of crop areas in nearly all counties of the United States can be downloaded from the Internet free of charge, whereas it is sometimes difficult to obtain any reliable data for a state or province (much coarser spatial units) in poorer countries where statistical bureaus lack adequate support. At a smaller geographic scale, much of the information on species distributions, crop yields, resource degradation, and so on is gathered from the most accessible areas near, for example, roads, research stations, and other human centers. The MA will need to carefully account for these biases at all scales, and may need to focus many analyses on regional and sub-regional case studies where adequate data are available.

Another type of bias arises from the tendency of scientists to collect data about "popular" taxa such as birds, mammals, butterflies, and trees at the expense of a more balanced coverage for all taxa (although the data coverage of even these popular taxa may suffer from geographic biases). Indeed, these popular taxa can be less important from the standpoint of ecosystem services than neglected groups such as bees, microbes, fungi, and aquatic plants. Not only are the spatial distributions of microbes and other such groups often poorly understood, but their ecological role in relation to ecosystem function and services is also not well documented. As a result, the MA may need to focus on case studies involving well-chosen indicator taxa as proxies and illustrative examples.

Other biases also have influenced the type of ecosystem data collected. For example, some types of data tend to be more abundant because they are easier to measure than others (point source wastewater discharges tend to be better documented in most watersheds, for instance, even though non-point discharges can also have a large influence on the state of water quality) or because they have a more direct effect on human welfare (for example, more data tend to be available about a river's impact on society during droughts and floods than under less catastrophic circumstances).

Differences in collection periods also present a challenge to data integration and quality. For instance, global initiatives on biodiversity assessment and soil degradation present different snapshots in time compiled from local data. While such data have their own quality problems, it is actually the change in such factors over time that is central to the concerns of the MA. But analyzing the change in ecosystems over time obviously requires time series data, which are often not available. Consequently, the MA may have to rely on only short-term trends, whose temporal resolution may or may not match those of the processes being studied. (See Chapter 5.)

The MA's integrated approach requires data on a wide variety of ecosystem services, their drivers, and their effects on human welfare. Yet the quality and coverage of these data vary greatly from one service to another. An example is the difference in data availability for provisioning versus supporting services of food production. The provisioning services are well described by abundant and relatively reliable data on crop and livestock production and on per capita food consumption. By comparison, the supporting services that make agricultural production possible, such as pollination and climate regulation, are much more poorly described. Nevertheless, for the sake of completeness, the MA must attempt to describe all aspects of ecosystem services, even those with poor data coverage.

Although current services of an ecosystem can be estimated, the MA must also determine whether these services can be sustained. But it is difficult and sometimes impossible to use current data to estimate the long-term sustainability of an ecosystem. As an example, it is possible to estimate the current production of a fishery, but nearly impossible to deduce from these data whether and for how long this production can be maintained. Hence we need information on the thresholds of sustainable production of natural resources. Sometimes this information can be provided by models that simulate the long-term dynamics of an ecosystem, as described later in this chapter.

In addition, assessing the contribution of ecosystem services to human well-being requires data that are usually not available. In particular, information is lacking on the material resources of individuals, their social relations, the state of governance, the role of freedoms and choices, and the state of equity. Moreover, available data are usually inadequate for analyzing temporal trends or for comparing one part of the world with another.

Another challenge for the MA is the use of traditional knowledge and undocumented experience. Because this information comes from sources outside of peer-reviewed publications, it needs to be critically assessed by other methods before being used. As an example, sometimes the quality of information (say, on the change in abundance over time of a particular plant or animal species) can be cross-checked from more than one source. Another step for controlling its quality would be to publicly archive the source and type of information. Archiving will also ensure that all researchers have access to this information.

Data Quality Assurance

Although quality assurance of data is obviously needed in any global assessment of world ecosystems and their services, there are different ways of achieving this. The method to be used within the MA has to take some special factors into account. First, global assessments typically rely on the voluntary efforts of numerous scientists and experts throughout the world. Second, the coordinators of global assessments normally do not have the capacity to examine carefully all the data sets to be used. In other words, the MA has neither the strong authority nor the capacity to intervene in the details of data analysis of its scientists. This does not mean that it should give up on quality control. On the contrary, the Intergovernmental Panel on Climate Change (IPCC) has shown that de facto quality control of data can be achieved without a formal quality assurance program. The following actions have made this possible:

- Most data used or cited by the IPCC stem from peer-reviewed scientific publications. It is expected that major deficiencies in data sources would be identified and "filtered out" in the course of peer-review.

- Some data sets come from large national or international organizations such as the United Nations Food and Agriculture Organization, the Center for International Earth Science Information Networks of Columbia University, or the United Nations Environment Programme (UNEP) World Conservation Monitoring Centre, that have internal procedures for maintaining quality control.

- Some data sets are assembled according to IPCC guidelines (such as emission inventories and estimates of carbon flux to forests). Data quality control is one of the aims of these guidelines.

To assure the quality of the data, the MA will build on the experience of the IPCC and insist on the use of data published in the scientific literature where possible. It will draw on the data of large organizations with their own data-control procedures and sponsor the development of its own data sets, as described earlier. Another step for quality control will be to set up a data archive containing metadata and some full data sets, as mentioned. This will give assessment coordinators an overview of much of the data being used in the assessment. Archiving would also help assure the quality of information coming from traditional knowledge and undocumented experience, as indicated earlier.

Indicator Selection

Global assessments of ecosystems and their services by definition involve the handling and evaluation of a huge number and variety of data and themes. It is clear that an assessment is only manageable if experts can focus on a limited number of representative indicators of ecosystems and their services. Because of the great weight these indicators hold, they must be carefully chosen. Earlier we described some of the particular types of indicators needed in an MA-type assessment. Here we pose the question, What are the characteristics of a "good indicator"? This depends on who is using the indicator and for what purpose, but three characteristics are common to all purposes: representativeness, reliability, and feasibility (Hardi and Zdan 1997; Prescott-Allen 2001).

For an indicator to be *representative*, it must cover the most important aspects of ecosystems and their services. As an example, consider the different possible indicators for "human health." "Life expectancy at birth" is not a bad indicator because it reflects all the causes of death that a typical person would be exposed to throughout life. "Healthy life expectancy at birth" is an even better indicator, however, because it subtracts the number of years likely to be lost to illness and injury.

For an indicator to be representative it must also be a sign of the degree to which an objective of an ecosystem service is met. For example, the indicator "healthy life expectancy at birth" shows the extent to which "having a long life in good health" has been attained, whereas immunization rates, health expenditures, and numbers of doctors are indirect indicators of this objective. Finally, to be representative, an indicator must illustrate trends in ecosystems and their services over time, as well as differences between places and groups of people.

An indicator is likely to be *reliable* if it is well founded, accurate, and measured in a standardized way using an established or peer-reviewed method and sound and consistent sampling procedures. And an indicator is *feasible* if it depends on data that are readily available or obtainable at reasonable cost.

The quality of potential indicators depends on how well they meet the above criteria. If no indicator can be found that adequately meets all these, then the component should be excluded from an assessment and its exclusion clearly noted.

The choice of components and indicators and their underlying methodologies must also be clearly documented. The more rigorous and systematic the choice of indicators, the more likely an indicator-based assessment will be transparent, consistent, and useful for decision-making. And the more involved decision-makers and stakeholders are in the selection of indicators, the greater will be their acceptance of results of the assessment. But a potential problem needs to be noted here: the time and technical skills required for selecting indicators might make it difficult for decision-makers and stakeholders to participate fully in the selection of indicators. This could work against the goal of maintaining an open process in the MA. (See Chapter 8.) At the same time, experts carrying out the assessment have the responsibility to ensure that the selection of indicators and the assessment as a whole are technically and scientifically sound. Hence in the area of indicators, as in other areas of the MA, a way must be found to maximize both the technical excellence of the assessment and the engagement of participants from government, civil society, and industry.

Units of Analysis and Reporting

Ecosystem Boundaries

Because the MA is concerned fundamentally with ecosystems and their functioning, it is necessary to describe these ecosystems and their spatial extent in as consistent a way possible, reflecting the state of scientific understanding. Indeed, many of the tasks described at the beginning of this chapter require an up-to-date characterization and mapping of the world's ecosystem types. For example, assessing the current state of ecosystems and their services or evaluating the impact of changes in these services on human well-being requires a consistent global overview of ecosystems.

At the most basic level, there are two fundamentally different ecosystem classifications: those based on actual ecosystem extent and those based on "original" or "potential" extent. The first type delineates ecosystem types based on their current distributions, including, for example, various agricultural and urban ecosystems developed by people through conversion of natural systems. The practical approach to assessing the location and extent of contemporary ecosystems at the regional scale has been through land cover interpretation of satellite data. For example, the International Geosphere-Biosphere Programme has identified 17 land cover types (deciduous broadleaf forest, for example, and cropland) using satellite data of 1-kilometer resolution (Belward 1996). These are widely accepted as proxies of aggregate ecosystem types.

Classifications of the second type, based on original or potential eco-system extent, attempt to depict the ecosystems that would occur without human modification—in other words, due to prevailing biotic and abiotic conditions. For example, the World Wide Fund for Nature has developed a global system of 871 terrestrial ecoregions, nested within 14 biomes and 8 biogeographic realms, based largely on patterns of potential natural veg-etation (Olson and Dinerstein 1998). Of course, marine ecosystems pro-vide special problems in defining ecosystem boundaries. Nevertheless, at least two marine classification systems exist, and they provide an estimate of boundaries between biogeochemical provinces and large marine eco-systems in the world's oceans (Longhurst 1991; Sherman and Duda 1999).

Ecosystem classifications of both types are likely to be useful to the MA. Current ecosystem data are essential for determining the services that eco-systems provide today as well as for establishing a baseline against which changes in land cover and services will be assessed using scenarios or in future assessments. At the same time, data on the original extent of ecosys-tems places patterns of land use change into ecological context. In fact, comparing the two types of classifications, especially where they differ, can yield insights into the relative extent of conversion of original habitat types.

Several issues must be considered. First, because factors defining eco-systems vary continuously in space, the boundaries of any set of ecologi-cally defined units will necessarily represent zones of transition instead of sharp boundaries. As a result, the precise location of these ecosystem bound-aries should be downplayed, and the meaning of the changes occurring across those lines emphasized.

Second, the appropriate ecosystem classification often will depend on the ecosystem service being considered. For example, in a mountainous region, analyses of fresh water would tend to link upland areas via stream and groundwater flow to the rest of the river basin below. Terrestrial analy-ses, in contrast, would link these same upland areas to areas of similar elevation on the other side of the divide, based on similarity of vegetation, fauna, and climate.

Third, ecosystem services operate at wide range of characteristic scales. (See Chapter 5.) Matching the scale of ecological assessment (and thus the units used) to the scale of the service considered will be an important, and often difficult, aspect of the MA's task.

Finally, even if ecosystems can be delineated with confidence, ecosys-tem processes and services often transcend local ecological units and bound-aries or involve interactions among them. For example, services provided by mangrove ecosystems (such as water purification, sediment capture, and habitat for juvenile fish) will be best maintained by proper management of both terrestrial and marine ecosystems. In addition, modern transportation systems have allowed ecosystems to provide services to people living far away, complicating and broadening the "ecological footprint" of human population centers.

Recognizing some of these difficulties in describing the location and extent of broad ecological systems, the MA has adopted definitions for such systems that allow for overlap in their extents. (See Chapter 2.) Thus areas of forest fragmented by patches opened up for agriculture are dealt with, from a systems perspective, in both the forest systems and the cultivated systems chapters of the Condition and Trends Reports, while cross-system summary tables control for possible double-counting of the ecosystem services provided.

Relating Ecological and Human-centered Units

An ecosystem's function and its ability to supply services to a particular human population are often best evaluated across its full extent, not only in the political unit in which that population lives. For example, water quality for a given municipality may depend more on the condition of the upstream portions of the watershed than on the areas within the city limits. At the same time, evaluating the importance of these ecosystem services to human welfare, as well as formulating policy to better manage them, will necessarily be conducted within the context of political units such as counties, cities, or provinces (Balvanera et al. 2001).

As a result, the MA conceptual framework will require frequent translation between ecological units and political or other society-centered units, particularly when linking indirect to direct drivers or ecosystem services to human well-being. For instance, demographic shifts may be an important indirect driver of many ecosystem changes, such as deforestation or soil erosion. Analyzing this relationship, however, will require relating demographic information collected for political units (such as counties) to ecological data necessarily assembled on ecological units (such as forest types). In addition, relating ecosystem services to human well-being, as in the water quality example, requires the reverse translation: from ecological units (watersheds) to political entities (cities).

Because ecological and political boundaries rarely overlap exactly, these translations among units are often difficult. For instance, it is hard to attribute human population densities collected on a national level to the country's ecosystems accurately.

Reporting Units

In order to best inform and assist the various users of MA products, it will be important to report assessment findings in units most relevant to those users. Many findings will be relevant to national and sub-national governments, and thus MA findings need to be reported in a form useful to these governments. In addition, the MA's scope and mandate clearly overlaps with those of existing international organizations and with previous scientific assessments (such as the Convention on Biological Diversity and the Convention to Combat Desertification). Hence, special efforts will be made to report MA findings in terms of the established units and frameworks used by these organizations.

Translating MA findings into various reporting units presents many challenges. In particular, the need to summarize the same findings in different forms has required careful collection and collation of information from the very start.

Modeling Issues

Models play an essential role in global assessments of ecosystems and their services. They can be used to analyze interactions between processes, fill data gaps, identify regions for data collection priority, or synthesize existing observations together into appropriate indicators of ecosystem services. They also provide the foundations for elaborating scenarios. As a result, models will play a synthesizing and integrative role in the MA, complementing the data collection and analysis efforts.

It is relevant to note that all models have built-in uncertainties linked to inaccurate or missing input data, weaknesses in driving forces, uncertain parameter values, simplified model structure, and other intrinsic model properties. One way of dealing with this uncertainty in the MA is to encourage the use of alternative models for computing the same ecosystem services and then compare model results. Having at least two independent sets of calculations can add confidence in the robustness of the model calculations, although it will not eliminate uncertainty.

To summarize the use of models in the MA, we have grouped them into two categories: environmental system models and human system models. The distinction between these two classes is somewhat blurred, however. What we call "environmental system models" often contain descriptions of some aspects of the human system, and "human system models" in turn often include aspects of environmental systems. Within each category, we identify some, although by no means all, classes of models that could be used in a global assessment.

Environmental System Models

A large number of freshwater resource models already exist and are used from local (small catchments), regional (watersheds and river networks), continental (large drainage basins), and global scales (e.g., Vörösmarty et al. 1989; Coe 2000; Donner et al. 2002; Alcamo et al. 2003). Included in this general class are the water balance and water transport models that consider the flow of water through plants, soil, underground, and storage systems. A new class of integrated water resource models expands these to include water use by society. These models can be used to assess how changes in a given component of the system affect the ability of other parts to provide ecosystem services associated with freshwater systems.

New models of marine resources are becoming available that can provide quantitative input to the assessment of ecosystem services provided by the marine environment. A representative of this group is the framework of models developed at the Fisheries Centre of the University of British Columbia (Walters et al. 1997; Pauly et al. 1998). Their approach, incorporated in the widely used Ecopath with Ecosim suite of software, is structured around a mass-balance concept that allows a simplified parameterization of the dynamics of freshwater and marine fisheries. These new models can be used to develop fisheries scenarios constrained by the feeding interactions within an ecosystem, thus leading to more realistic scenarios than the traditional fisheries management approaches, where such constraints are ignored. The modeling framework of the University of British Columbia also depicts fishery dynamics on a spatial grid of the world's oceans, thus providing a spatially explicit estimate of changes in the ecosystem services associated with the world's marine resources.

There are numerous models of terrestrial ecosystem processes that are appropriate for analyses at the local, regional, and global scale (Prentice et al. 1992; Melillo et al. 1993; Alcamo et al. 1994; Foley et al. 1996; Kucharik et al. 2000). Biogeochemistry models describe the flow of energy, water, and nutrients in the biosphere and are used to estimate essential properties such as productivity, carbon storage, and other functional aspects of ecosystems. At a more general level, biogeography models are used to describe patterns of plant distribution with respect to climate and soils and can be used to test the impact of changes in those variables. Land cover models provide insights in land cover change by analyzing the relationship between the various drivers of the process; such models are often spatially explicit and can help in assessing the impact of decisions affecting the use of land. Finally, integrated global ecosystem models provide a dynamic framework for studying changes in ecosystem structure and function under changing pressures. These models have largely focused on natural vegetation systems but are starting to turn to managed ecosystems.

A wide variety of climate models exists, and some of them can be used to quantify relationships between ecosystems and climate (Cox et al. 2000; Foley et al. 2000; Wang and Eltahir 2000). In particular, they help in examining both how ecosystems contribute to climate regulation and, conversely, how changes in climate may affect the capacity of ecosystems to provide goods and services in the future. General circulation models (GCMs) have been the traditional working tool for climate research, but up to now their linkage with ecosystems has been limited mostly to their representation of the influence of surface albedo on energy fluxes. Fully coupled climate-biosphere models are extensions of GCMs; they simulate physical and biogeochemical interactions between ecosystems and the climate system. These models, which can be of varying complexity, are more relevant to the MA.

For the most part, each genre of environmental models can be applied at various scales—local, regional, continental, and global. Their usefulness at various scales depends on their capability to capture input data and processes at a resolution that is consistent with processes at play at those scales. At local scales, models may be used to demonstrate the characteristic dynamics of ecosystems in different geographic areas where observational data are lacking. At regional and continental scales, models can assist in making up for observational data deficiencies and addressing biome-wide issues. At the global scale, models could be used to describe, among other subjects, changes in vegetation cover and biodiversity, linkages between global hydrology and water use, and food and crop production in a changing economic or climatic context. They also provide a standardized method for computing ecosystem indicators everywhere in the world. (See also Chapter 5.)

Human System Models

Social scientists model human behavior at various levels of aggregation, such as at the household level, the sub-national sectoral level, and the national and international level. Although these models strive toward quantification, purely conceptual models also play an important role in social science thinking and policy decision-making.

Household models examine the impact of changes in the external environment on production, consumption, and investment decisions. They have been used in particular to analyze differences between households in their access to resources. By comparison, sectoral models describe the various components of a complete economic sector. Sectoral models are used to address questions about the relationship between external factors and the performance of the sector—for example, anticipating the impact of a falling global wheat price on wheat production in Asia. Recently researchers have begun to apply sectoral models to the question of the impact of a particular economic sector on natural resources, as in the case of the impact of agricultural production on the availability of land and water (and vice versa) (e.g., Rosegrant et al. 2002).

Some human system models, particularly economic models, are available at the national and international level. They describe either a particular sector (for instance, energy or agriculture) at this level or a grouping of sectors. A particular class of national and global models is made up of the computable general equilibrium (CGE) models, which trace through economy-wide linkages of changes that are targeted to particular sectors. CGE models have the potential to be used for assessing the consequences of environmental change, but few examples of such use exist.

It should be stressed that the majority of "human system models" focus on economic efficiency and the economically optimal use of natural resources. Thus the broader issues of human well-being addressed by the MA, including such factors as freedom of choice, security, and health, will require a new generation of models. At a minimum, the present cadre of models needs to be extended to address these critical constituents of human well-being and their links to ecosystem services.

Integrated Models

There is also a small set of global integrated models that combine descriptions of the environmental system with the human system (e.g., Alcamo et al. 1996; Edmonds et al. 1996; Kainuma et al. 2003). These models relate demographic, economic, and technological factors with global changes in climate, natural vegetation, agricultural production, water resources, and other aspects of the Earth system. Some take into account feedbacks from the environmental system to the human system. Such models can be used in the MA to fill in data gaps in describing the current state of ecosystem services and for generating scenarios of future ecosystem services.

Scenario Analysis

The MA is concerned not only with assessing the current state and historical trends of ecosystems but also with developing medium- and long-term scenarios. This is because decision-making involves not only immediate options but also their consequences for the future (Carpenter 2002). Known or potentially long-lasting effects (decades or longer) must be explicitly taken into account in the decision-making process. Of particular relevance are situations where management decisions lead to irreversible changes in ecosystem conditions and processes. In these cases the policy decision must be informed by the probability of reaching such an irreversible threshold in the set time frame.

Ecology has many methods for anticipating the future of ecosystems (Clark et al. 2000). These include prediction, forecasting, and projecting, each with its own methods for estimating ecological outcomes, probabilities, and uncertainties. Ecological forecasts themselves are insufficient for the needs of the MA, however. (See Box 7.1.) Although the MA will use forecasts and other types of model projections where possible, additional methods are needed to provide a more comprehensive coverage of future ecological change in a format useful for decision-making. Scenarios are one of those alternatives.

Scenarios for Ecological Services

The MA will use scenarios to summarize and communicate the diverse trajectories that the world's ecosystems may take in future decades. Scenarios are plausible alternative futures, each an example of what might happen under particular assumptions. They reveal the dynamic processes and causal chains leading to different outcomes of the future (Rotmans et al. 2000). Scenarios can be used as a systematic method for thinking creatively about complex, uncertain futures. In this way, they help us to understand the upcoming choices that need to be made and highlight developments in the present (Rotmans et al. 2000). In our case, we are particularly concerned with scenarios that deal with changes in ecosystem services and their impact on human well-being.

The MA Scenarios Working Group will develop scenarios that connect possible changes in drivers (which may be unpredictable or uncontrollable) with human demands for ecosystem services. The scenarios will link these demands, in turn, to the futures of the services themselves and the aspects of human welfare that depend on them. The scenario building exercise will break new ground in several areas:

BOX 7.1   Ecological Forecasting

While ecological forecasting has had notable success in a limited number of well-studied cases (Clark et al. 2000; Carpenter 2002), scientists' ability to forecast ecological change and its probability distributions has important limitations. Often the amount of information available for projecting ecosystem behavior is insufficient. Some particularly large changes in ecosystems occur only infrequently and are therefore difficult to study, characterize, and predict (Turner and Dale 1998). Other changes are simply random. The dynamics of socioecological systems are especially challenging, and most of the systems of interest to the Millennium Ecosystem Assessment are socioecological ones. Last, many of the current and anticipated changes in ecosystems, and in human use of ecosystems, are new, and there is therefore no historical experience on which to base forecasts.

   For these reasons, the probability distributions of ecological predictions or forecasts frequently cannot be characterized (Ludwig et al. 2001; Carpenter 2002). Ecological forecasts may also have many dimensions or contingencies, which means that a large number of potential outcomes must be considered. The multiplicity, contingency, and complexity of these many potential outcomes may be a barrier to understanding that limits the usefulness of the forecasts for decision-makers or the general public.

- development of scenarios for global futures linked explicitly to ecosystem services and the human consequences of ecosystem change,

- consideration of trade-offs among individual ecosystem services within the "bundle" of benefits that any particular ecosystem potentially provides to society,

- assessment of modeling capabilities for linking socioeconomic drivers and ecosystem services, and

- consideration of ambiguous futures as well as quantifiable uncertainties.

Review of Scenario Types and Approaches

Scenario analysis was first used for strategic planning during the early cold war period. However, scenarios about long-term sustainability of natural resource use did not emerge until the 1970s. These studies included the well-known report by Meadows et al. (1972) in which the authors discussed limits to human population growth. Scenarios were also being used by some businesses at this time, including Royal Dutch/Shell (Wack 1985), that have since become leaders in the field of scenario use for business and other uses.

Since 1995, there has been widespread use of scenarios to assess the status of the global environment. The MA intends to build on these examples, such as the reports of the Global Scenarios Group, UNEP's *Global Environmental Outlook*, the *Special Report on Emissions Scenarios* released by the IPCC, the scenarios of the World Business Council on Sustainable Development, the *World Water Vision Scenarios* of the World Water Commission, and the scenarios computed with the IMAGE model, to explore long-range dynamics of global environmental change. (See Table 7.1.)

In general, scenarios contain a description of step-wise changes, driving forces, base year, time horizon and time steps, and a storyline (Alcamo 2001). They are often classified by the method used to develop them, the goals and objectives, or the output. One classification of scenarios discriminates between "exploratory" and "anticipatory" scenarios. Exploratory scenarios are descriptive: they begin in the present and explore trends into the future. Anticipatory scenarios start with a vision of the future that could be optimistic, pessimistic, or neutral and work backwards in time to imagine how society might reach that future. The MA approach to development scenarios is likely to be a mixture of exploratory and anticipatory approaches.

Scenarios can be built around qualitative information, quantitative information, or a combination of both. Qualitative scenarios include qualitative information and use a narrative text to convey the main scenario messages. This can be helpful when presenting information to a nonscientific audience. Quantitative scenarios usually rely on models based on quan-

TABLE 7.1   Overview of Some Global Scenario Projects

Name	Description	Citation
Global Scenario Group (GSG)	Examines global scenarios based on three classes: conventional worlds, barbarization, and great transitions	Gallopin 1997, Raskin et al. 1998, Raskin et al. 2002
Global Environmental Outlook 3 (GEO-3)	Similar to GSG, with emphasis on regional texture	UNEP 2002
World Business Council on Sustainable Development (WBCSD)	Scenarios aimed at helping corporate members reflect on the business risks and opportunities of the sustainable development challenge (FROG!, GEOpolity, and Jazz)	WBCSD 1997
World Water Vision (WWV)	Three global water scenarios focusing on water supply and demand, including water requirements for ecosystems	Cosgrove and Rijsberman 2000, Gallopin and Rijsberman 2000
IPCC Special Report on Emission Scenarios (SRES)	Greenhouse gas emissions scenarios to the year 2100; axes of change are sustainable to unsustainable, and globally integrated to globally fragmented	SRES 2000

titative information to calculate future developments and changes; they are presented in the form of graphs and tables (Alcamo 2001). Both scenario types can be combined to develop internally consistent storylines based on quantification with models, which are then disseminated in a narrative form. This approach will be used to develop the MA scenarios. That is, we will develop a general qualitative storyline supported by quantification. Scenario development will be an iterative process, involving development of zero-order storylines, quantification of driving forces and indicators, and revision of the storylines together with various scenario user groups.

According to Alcamo (2001), good scenarios fulfill the objectives of the exercise; are sufficiently documented; are plausible; are internally consistent; challenge the beliefs and broaden the understanding of readers (experts, policy-makers, and laypeople); and convey complex interactions in the socioecological system. We will attempt to meet these goals through a participatory process that involves dialogue among scenario experts, scientists, decision-makers, user communities, and others.

The MA Approach to Scenario Analysis

At the most general level, the MA scenarios should connect possible changes in drivers with human demands for ecosystem services and, in turn, to the futures of the ecosystem services themselves and the aspects of human welfare that depend on them. This is a complex task.

Some of the drivers that might be considered ambiguous and uncontrollable include governance, economic globalization, climate, or emergence of disease. For example, the MA scenarios could consider the implications of increasing interconnectedness of economies at the global scale. How will such global economic changes affect the capacity of ecosystems to produce food and fiber, provide fresh water, and sustain biodiversity? What are the impacts of these ecological changes for the alleviation of poverty? And what are the implications for ecosystem services of changes in human welfare? Such feedbacks are at the heart of MA scenarios.

The Scenarios Working Group developed the following objectives to guide its scenario-building work:

- to illustrate that global changes are connected to ecosystem services at every scale, from global to local, and that these changes have implications for human well-being;
- to highlight major trade-offs among ecosystem services;

- to illustrate the effectiveness of different policies in making ecosystem services available and maintaining these services, including evaluating the effectiveness of policies at different scales; and
- to fulfill the objectives of scenario users.

The objective of the scenario-building exercise can also be summed up by the question, What are the possible co-evolutions of humanity and Earth's ecosystems? Several other more specific questions follow logically from this first one:

- How will ecosystem services support human well-being in the future?
- What are the major threats to the world's ecosystems?
- What are the trade-offs (in space, between current and future use, between ecosystem services, and so on)?
- What can be done to harmonize human welfare and production of ecosystem services?
- What are the appropriate incentive structures to ensure that ecosystems are used wisely?
- What are the signatures of different drivers of ecosystem goods and services and human well-being?
- What are the threats and opportunities for provision of ecosystem services?
- What are the appropriate scales for addressing ecosystem services, drivers, and interventions?

The current proposal under consideration by the Scenarios Working Group is to develop four or five scenarios. The group first evaluated five "zero-order" (very preliminary) scenarios found in previous global scenario exercises. (See Table 7.2.) Although the previous scenarios are detailed and carefully constructed, their focus is largely on social and economic issues. Environmental changes enter into many of them, both directly (for example, in the IPCC scenarios on global climate change) and indirectly (as drivers of societal change, for instance), but the many complex feedbacks that characterize real ecosystems are not explored or tested in detail in any existing global scenario.

The MA will approach the construction of global scenarios from an explicitly ecological perspective. That is, we will draw on previous scenarios but will focus on ecological surprises and cross-scale ecological feedbacks. MA scenarios should address branch points in global dynamics that are related to changes in ecosystem services. For example, how would the

TABLE 7.2 **Zero-order Millennium Ecosystem Assessment Storylines Derived from Previous Global Scenario Exercises**

Name*	Key Words	Similar To
EGS-1	Market-driven globalization, trade liberalization, institutional modernization	IPCC: A1 GEO-3: markets first GSG: market forces
EGS-2	As above, except strong policy focus on sustainability	IPCC: B1 GEO-3: market first / policy first GSG: market forces + policy reform
EGS-3	Value shift toward sustainability in industrial world; policy focus on poverty, sustainability	IPCC: B1 GSG: great transition GEO-3: sustainability first
EGS-4	Fragmented development; conservation of local identities; regionalization of economies	IPCC: A2, B2 GSG: multiworlds
EGS-5	Elites in fortresses (national or local); poverty and repression outside	WWV: business as usual GSG: fortress world GEO-3: security first

* EGS = ecosystem global scenario

global system change if ecosystems are more fragile than expected, or more robust than expected?

Scenarios will be developed for the global system. Quantitative outputs of the scenarios will be aggregated from regional data. As with previous global scenarios, a regional breakdown of quantitative outputs will be provided in some cases. Quantification will be accomplished using a combination of the models developed for other global scenarios projects, as described in this section. (See Table 7.3.)

Indicators will be chosen so that they reflect user needs, integrate information across ecosystem types, connect clearly to human well-being, are compelling, have scientific legitimacy, and are scalable. They should also be useful in estimating the vulnerability of society to changes in ecosystem services, including society's ability to cope and adapt to these changes.

Models to Support Scenario Analysis

As noted, part of the MA strategy for scenario analysis calls for the use of models to "quantify" the scenarios—that is, to generate quantitative aspects of the scenarios. For this task a wide range of models will be needed, as large a variety as described earlier for filling in data gaps.

Models will be used to "translate" the language of the scenarios into quantitative illustrations of changes in ecosystem services. The family of

TABLE 7.3   Matching of Millennium Ecosystem Assessment Scenarios with Earlier Scenario Exercises

Earlier Exercises	Models	EGS-1	EGS-2	EGS-3	EGS-4	EGS-5
GSG	PoleStar	Market forces	Policy reform	Great transitions	Eco-communalism	Fortress world
SRES	AIM, IMAGE, MESSAGE, MARIA, MINICAM, ASF	A1	A1-policy, B1	B1-policy	B2/A2-policy	A2
GEO-3	PoleStar, IMAGE, AIM, WaterGap Globio	Markets first	Policy first	Sustainability first	—	Security first
WWV	PoleStar, WaterGap IFPRI	TEC	TEC	VAL	—	BaU
WBCSD	—	FROG!	GEOpolity	Jazz	—	—
OECD	Jobs, PoleStar	Reference	Policy variants			

scenarios will each have associated changes in indirect and direct drivers—these can be used to drive process-based models of ecosystem services to help determine the ecological outcomes of the scenarios. For example, changes in climate, land use patterns, and water demand may be fed into watershed models to assess changes in freshwater availability, water quality, and aquatic habitats. Likewise, changes in forest cover and climate could be used to drive models of habitat loss in order to assess changes in biological diversity.

Because the MA is a multiscale assessment, and because the scenarios will be evaluated at multiple scales, modeling will be performed at local, regional, and global scales. At the global scale, gross changes in ecosystem services may be responding to changes in climate, atmospheric chemistry, and patterns of land use. Such modeling exercises could help pinpoint changes in freshwater availability, crop production, carbon sequestration, and habitat. At regional scales, modeling exercises could help illustrate more detailed outcomes of the scenarios: changes in water flows, agricultural systems, disease pathways, and water quality may be addressed at these scales. Finally, at local scales, questions related to community access to natural resources, as well as the relationships between environmental conditions and human health, may be best addressed.

Ultimately, models provide the means of translating the storylines of scenarios into quantitative assessments of changing ecosystem services. The degree of quantification that is performed will likely be somewhat limited in scope, as models are not available for every ecosystem process at every scale.

Overarching Issues

Matters of Scale

The issue of scale arises in nearly all aspects of an MA-type of assessment. By the "scale issue" we mean the question of whether data analyses and data comparisons correctly take into account the different aggregation levels by which ecosystems can be described. Here we only mention some of the main points of this issue, as Chapter 5 covers these questions more completely.

The scale issue is critical to the analytical approach of the MA because ecosystems operate and are measured and observed at different scales. At each scale researchers characterize the extent, pressures, conditions, and trends of ecosystem types. For any size patch other than the global scale, there will be a set of factors external to the ecosystem that influence how it functions and, in turn, there will be flows of mass and energy between the patch and the larger scales. On one hand, the larger the scale, the more inclusive the description of mass and energy flows. On the other hand, the larger the scale, the rougher the description of the ecosystem. Hence part of the scale issue is determining the correct spatial and temporal coverage and resolution to assess ecosystems and their services and drivers. Other examples of scale issues that must be addressed by the MA include the following:

- There needs to be as close a match as possible between the scale used to map ecosystems and the scale required to characterize ecosystem services.

- Ecosystem services themselves are described at different scales. For example, some services (such as providing fresh water) tend to operate more locally than others (such as climate regulation). The differences in scales must be taken into account in comparing the value of different ecosystem services.

- Many scale issues arise when models are used to provide information for an assessment. For example, coarse-scaled output from global climate models may be difficult to apply to local decisions or to use as input to finer-scaled local vegetation models.

- The analysis of response options also raises complex issues of scale. Often the management of natural resources such as forests or fisheries involves many different political and economic actors (local and na-

tional governments, for instance, and local and multinational companies) operating at many different spatial and organizational scales.

Review and Validation Procedures

The MA assessment reports will undergo two rounds of peer-review involving experts and governments. An independent Review Board has also
been established to oversee this review process and to ensure that the review comments received are handled appropriately by the assessment authors. Much of the information contained in the assessment reports will
be based on published scientific literature, which in turn has been through
a formal process of peer review. However, the MA also seeks to incorporate information from traditional knowledge, practitioners' knowledge, and
undocumented experience. This is particularly important in the case of
the MA sub-global assessments—particularly the community-scale ones—
since much of the information available for these may not be in the form
of published scientific articles. Each of the MA sub-global assessments
will develop a process to validate unpublished information, including many,
if not all, of the following features:

- self-critical review notes or reflective diaries—the researcher should
  record information on his or her own perceptions of where information
  being recorded may be incomplete, biased, or in error;

- triangulation—multiple sources of information should be obtained, particularly for critical pieces of information;

- review by communities—where the information involves local or traditional knowledge, members of the community should be given an
  opportunity to review the findings prior to finalization of the assessment; and

- review by stakeholders at higher and lower scales—individuals who may
  not have detailed local knowledge of the area being assessed, but with
  knowledge of the region in which the assessment is located, should be
  given an opportunity to review the findings prior to finalization of the
  assessment.

In addition, when unpublished information is included in the global
MA assessment reports, detailed information concerning the source of the
information (such as names of people interviewed, dates and types of notes
recorded, the presence or absence of a researcher's self-critical review notes,
and other sources of information validating the information) will be made
available to the co-chairs of the Working Group.

Analysis of Uncertainty

This section draws heavily on the document developed for handling uncertainty in IPPC assessments (Moss and Schneider 2000).

An assessment of the relative credibility of the range of ecosystem conditions, processes, and outcomes should be a major goal of assessment reports. It is important to adopt a consistent approach for assessing, characterizing, and reporting uncertainties. This will help improve communication between the research community and decision-makers regarding what is known and unknown (and to what degree) about the relevant issues covered in the assessment.

The scientific community must bear in mind that users of assessment reports are likely to estimate for themselves the extent of uncertainties if authors do not provide uncertainty estimates. Hence it is desirable for experts to give their best estimates of these uncertainties (e.g., Morgan and Henrion 1990).

An "uncertain estimate" can mean different things to different experts, ranging from an estimate just short of complete certainty to an informed guess or speculation. Sometimes uncertainty results from a lack of information; on other occasions it is caused by disagreement about what is known or even knowable. Some categories of uncertainty are amenable to quantification, while other kinds cannot be sensibly expressed in terms of probabilities. (See Schneider et al. 1998 for a survey of the literature on characterizations of uncertainty.)

Uncertainty is not unique to the domains of biophysical and socioeconomic research. Uncertainties also arise from such factors as linguistic imprecision, statistical variation, measurement error, variability, approximation, subjective judgment, and disagreement. These problems can be compounded, however, by additional characteristics of environmental change research, such as potentially long time lags between driving forces and response at larger scales. Moreover, because environmental change and other complex, sociotechnical policy issues are not just scientific topics but also matters of public debate, it is important to recognize that even good data and thoughtful analysis may be insufficient to dispel some aspects of uncertainty associated with the different standards of evidence and degrees of risk aversion or acceptance that individuals may hold (Morgan 1998; Casman et al. 1999).

In many cases, a "Bayesian" or "subjective" characterization of probability will be appropriate (Gelman et al. 1995; Bernardo and Smith 2000). The Bayesian paradigm is a formal and rigorous method for calculating probabilities, and is often used in the "rational" analysis of decisions

(Lindley 1985; Pratt et al. 1995). Bayesian statistics can be used to calculate probability distributions in the absence of information by using prior distributions that represent best estimates by the scientists making the calculations. This is a different type of subjectivity, which must be addressed in a straightforward and transparent way in the MA calculations.

Although "science" itself strives for objective empirical information to test theory and models, "science for policy" must be recognized as a different enterprise, involving being responsive to policy-makers' needs for expert judgment at a particular time, given the information currently available, even if those judgments involve a considerable degree of subjectivity. Such subjectivity should be both consistently expressed (linked to quantitative distributions when possible), and explicitly stated so that well-established and highly subjective judgments are less likely to get confounded in policy debates. The key point is that authors should explicitly state what sort of approach they are using in a particular case. Transparency is the key in all cases.

Vague or broad statements of "medium confidence" that are difficult to support or refute should be avoided. For example, scientists could have at least medium confidence that "desalinization could alter biodiversity." Such a statement is not particularly informative unless the degree of desalinization and the direction and severity of the biodiversity change are specified. The point is to avoid conclusions that are essentially indifferent statements based on speculative knowledge.

The procedure for carrying out an uncertainty analysis depends very much on the data and information available about a particular subject. Where the amount of information is relatively rich, the following procedure can be followed:

- For each major finding, identify the most important factors and uncertainties that are likely to affect the conclusions.

- Document ranges and distributions from the literature, including sources of information on the key causes of uncertainty and the types of evidence available to support a finding.

- Make an initial determination of the appropriate level of precision— determine whether quantitative estimates are possible, or only qualitative statements.

- Specify the distribution of values that a parameter, variable, or outcome may take in either quantitative or qualitative form. Identify end points of the range and provide an assessment of the central tendency and general shape of the distribution, if appropriate.

- Rate and describe the state of scientific information on which the conclusions or estimates in the preceding step are based.
- Prepare a "traceable account" of how the estimates were constructed, describing reasons for adopting a particular probability distribution.

Note that some of these steps (particularly those having to do with estimating the probability distributions of parameters and variables) sometimes must be omitted because of lack of information or time to carry out a full analysis.

Not only is the method for assessing uncertainty important, so is the communication of uncertainty. Among the effective ways to communicate uncertainty is to present it in clear graphical form. Various approaches for graphical presentation of uncertainties are available, involving trade-offs between simplicity and sophistication, particularly in the choice of the number of dimensions to use in presenting the information. Using various approaches, the degree to which experts agree on the uncertainty estimates can also be depicted.

Conclusion

The aim of this chapter has been to provide a road-map for how the MA will be carried out. We have pointed out that such a complex and comprehensive assessment will raise many difficult issues about data handling, data analysis, uses of modeling, scenario analysis, and so on. Although some of these issues will only be resolved in the course of implementing the MA, this chapter suggests many useful actions for resolving these issues. Taken together, these actions make up a coherent analytical approach for achieving the goals of the Millennium Ecosystem Assessment.

8 Strategic Interventions, Response Options, and Decision-making

EXECUTIVE SUMMARY

- Decision-making processes and institutions operate across spatial scales and organizational levels—from the village to the planet. Decision processes are value-based and combine political and technical elements to varying degrees. Desirable properties of decision-making processes include equity, attention to vulnerability, transparency, accountability, and participation.

- Strategies and interventions that will help meet societies' goals for the conservation and sustainable use of ecosystems include incorporating the value of ecosystems in decisions, channeling diffuse ecosystem benefits to decision-makers with focused local interests, creating markets and property rights, educating and dispersing knowledge, and investing to improve ecosystems and the services they provide.

- The choice among options will be greatly influenced by the temporal and physical scale of the problem or opportunity, the uncertainties, the cultural context, and questions of equity.

- Mechanisms for accomplishing these interventions include conventions, laws, regulations, and enforcement; contracts, partnerships, and collaboration; and private and public action.

- Institutions at different levels have different response options available to them, and special care is required to ensure policy coherence. Decision-making processes combine problem identification and analysis, policy option identification, policy choice, policy implementation, and monitoring and evaluation in an iterative fashion.

- A range of tools is available to choose among response options—from cultural prescriptive rules to cost-benefit and cost-effectiveness analysis. In the selection of an analytical tool and in the evaluation of response options, the social, economic, environmental, and historical context should be taken into account.

- Policies at each level and scale need to be adaptive and flexible in order to learn from past experience, to hedge against risk, and to consider uncertainty. However, trade-offs between the responsiveness and the stability of the policy environment need to be considered.

■ Intermediate indicators may be required to link policies and actions and their impacts on ecosystems and human well-being. Quantitative indicators make the trade-offs in policy-making explicit, but qualitative information is valuable where measurement is not possible. Traditional and practitioner knowledge are important sources in addition to science.

■ "Boundary organizations" that synthesize and translate scientific research and explore the policy implications can bridge the gap between science and decision-making. Journalists have a similar bridging responsibility to ensure that science and policy information is transmitted to the public in ways that are both objective and engaging.

Introduction

The context of decision-making about ecosystems is changing rapidly. World population continues to grow and become more urban, consumption is increasing, the climate is changing, and human actions increasingly influence major biogeochemical cycles and the majority of ecosystems. In addition, the ecosystems people depend on for services are more tightly coupled to each other and to human systems and in many cases are more stressed. At the same time, however, scientists and others are developing a far better understanding of how ecosystems function, how they generate ecosystem services, how those services may contribute to human welfare, and how values can be assigned to the services.

Thus the new challenge to decision-making is to make effective use of new information and tools in this changing context in order to improve the decisions that intend to enhance human well-being and provide for a sustainable flow of ecosystem services. It seems clear that the choices of the past may not be the most appropriate strategy for the future, and that even the way people think about intervening in ecosystems must be revised to take account of new information, new tools, and new contexts. In addition, some old challenges must still be addressed.

Perhaps the most important traditional challenge is the complex trade-off faced when making decisions about how to alter ecosystems with the goal of enhancing the flow of services. Increasing the flow of one service from a system, such as provision of timber, may decrease the flow from others, such as carbon sequestration or the provision of habitat. In addition, benefits, costs, and risk are not allocated equally to everyone, so any intervention will change the distribution of human well-being—another trade-off.

These trade-offs are related to a second ongoing problem: some benefits of ecosystem services are easily captured by those who have access to the system, while others are harder to capture locally. For example, it may be relatively easy for local people to capture the direct use value of timber in a forest via market prices—they are capturing the value of provisioning services. At the same time, people around the world may benefit from carbon sequestration by the forest—an indirect use value of a supporting service. Under many institutional arrangements, the people near the forest have no way to capture this other value. Further, some cultural services of ecosystems and the existence value of biodiversity are global and thus difficult for local people to capture. Because the direct use value— revenues from logging—can easily be converted into income for local people, for local and national governments, and for local, regional, and multinational firms, there is a strong incentive to log the forest. In contrast, the indirect use and existence value—carbon sequestration and appreciation of old-growth forests—are much harder to translate into income for anyone. As a result, there will be a tendency for decisions to favor the direct use even though a full analysis of the total value of ecosystem services might favor preserving or enhancing the indirect use and existence values by not logging.

The characteristics of the ecosystem, the technologies available for using it and monitoring such use, and the institutional arrangements that distribute values across groups all have consequences for what decisions are made (Ostrom et al. 1999; Dietz et al. 2002b). A great deal is understood about these problems, and the state of the science often provides guidance on the design of institutions to promote capturing the full value of an ecosystem (Costanza and Folke 1996; Stern et al. 2002).

Decision-making Processes

The Millennium Ecosystem Assessment (MA) must look carefully at decision-making processes for choosing among alternative intervention strategies. Process influences the intervention chosen. It can also influence those who bring about or respond to an intervention and who facilitate or retard the ability to adapt to changing circumstances. Of course, decision-making processes vary across jurisdictions, institutions, and cultures. But broadly accepted norms regarding decision-making and analyses of how decision processes handle information and influence implementation (Hemmati 2001; Petkova et al. 2002; Dietz 2003) suggest some desirable characteristics regarding:

- use of the best information,

- transparency and participation,
- equity and vulnerability,
- cognitive and organizational strengths and weaknesses,
- lessons from past decisions and the protection of options,
- accountability,
- efficiency, and
- cumulative and cross-scale effects.

The MA is motivated by recent improvements in information about ecosystems, the services they provide, the impact of those services on human well-being, the value of those services, and the design of institutions, programs, and policies to shape behavior. In addition, new tools to use that understanding are being developed. Current decision-making practices often do not reflect these important developments. For example, relatively few decisions take account of indirect use value and very few take explicit account of existence values. As a result, many decisions about intervention into ecosystems are not based on the best possible information. Note that information about both facts and values is required and that information used to make decisions about ecosystems will always be uncertain and involve risk. Thus knowledge about uncertainty and risk is itself an important component of the decision-making process, as discussed later in the chapter.

Processes that are transparent and that involve all those who will be affected by the decision are more likely to be seen as legitimate and to find support when implemented (U.S. National Research Council 1999; U.S. EPA Science Advisory Board 2000). Further, the management of ecosystems requires locally grounded knowledge (often referred to as "traditional ecological knowledge") and must address questions that can emerge only from an understanding of local situations (Stern and Fineberg 1996; Dietz and Stern 1998; Berkes 2002). That knowledge can be obtained only by interaction with those who have local experience. Finally, since non-use values are an important contribution of many ecosystems to human welfare, people who are not local to an ecosystem but who benefit from its non-use values must also be engaged. This implies that decision-making processes should involve stakeholders effectively, a principle that has become central to risk analysis.

Although no universal prescriptions on how best to do this are possible, a growing literature on public participation in environmental decision-making provides useful guidance (Stanner 1979; Fiorino 1990; Dietz

1994; Renn et al. 1995; Slocum et al. 1995; Stern and Fineberg 1996; Chess et al. 1998; Chess and Purcell 1999; Webler 1999; Beierle and Cayford 2002). Sometimes formal negotiation and conflict resolution processes are helpful, but in almost all cases careful design of participation mechanisms is important to good decision-making.

In terms of equity and vulnerability, changing the provision of ecosystem services very often produces "winners" and "losers." For effective implementation, the benefits, costs, and risks across groups must be balanced in an equitable way (Agrawal 2002; McCay 2002). Given that many changes in ecosystems can have important impacts on the poor, special attention to the most vulnerable populations is also warranted, as is special attention to human health.

Individuals, groups, communities, and organizations have varying strengths and weaknesses in processing information (Kahneman et al. 1982; but see Cosmides and Tooby 1996; Wilson 2002). Decision processes will be most effective if they make use of the kinds of decision tools described in this chapter to compensate for limits and weaknesses.

The understanding of ecosystem dynamics will always be limited, socioeconomic systems will continue to change, and outside determinants can never be fully anticipated. As Campbell (1969) noted over three decades ago, all policies are experiments. Decisions should consider whether or not a course of action is reversible and should incorporate, whenever possible, procedures to evaluate the outcomes of actions and learn from them. That is, people should try to learn from these experiments and use that knowledge in designing new ones. Debate about how exactly to do this continues in discussions of adaptive management, social learning, safe minimum standards, and the precautionary principle (Gunderson et al. 1995b; Yohe and Toth 2000). But the core message of all approaches is the same: acknowledge the limits of human understanding, give special consideration to irreversible changes, and evaluate the impacts of decisions as they unfold.

In terms of accountability, the consequences of decisions do not always redound directly to those who make them. As noted earlier, those who might decide to harvest timber from a forest may not bear any of the consequences of disrupting the flow of supporting and cultural services and so will not take such services into consideration in making their decision. This problem is exacerbated in the face of uncertainty and risk—the relationship between a decision and its consequences is hard to see. Effective decision-making can develop only if the people making decisions are accountable for the results (Perrow 1984). Unfortunately, in many circumstances the lack of accountability removes the incentive for decision-makers to use the best information available.

In a world of scarce resources—fiscal, human, and natural—efficiency should be an important criterion for choosing among intervention options. This is a central tenet of environmental and resource economics, and there are numerous effective tools for examining the efficiency of various options, as described later. In reality, the goals of equity, encouraging learning, and protecting options need to be considered together with the goal of maximizing efficiency. This typically leads to a multicriteria decision problem.

Many decisions about interventions into ecosystems are made at a local level. As noted earlier, this involves balancing locally concentrated costs and benefits against those that are more widely distributed and harder to capture. Another way to think about this is that decisions based only on a local analysis can miss cumulative effects of the same kind of decision being taken in multiple localities. Thus, too narrow a scope of analysis results in decisions that are less than optimal from a larger perspective (Olson 1965). Appropriate decisions emerge only when all relevant scales are considered.

Although these eight features of decision processes seem consequential and deserve serious attention, it is unclear exactly how they influence decisions and implementation and especially how the impact varies across contexts. Analysis of how the characteristics of the decision process influence changes in ecosystem services and human well-being deserves careful attention in the MA.

Response Options and Strategic Interventions

There are many options for responding to the need to protect and restore ecosystems and the services they provide and to ensure the equitable distribution of the benefits of those services. Fundamentally, these options can be characterized as interventions that stimulate or suppress certain human activities and those that create knowledge or investment. They can take the form of prescriptions of behavior (that is, "command and control" or assignment of accountabilities), incentives and disincentives (that is, creating or assigning property rights or establishing markets, subsidies, and taxes), education and knowledge sharing, or direct investment and expenditure (Kaufmann-Hayoz et al. 2001; Dietz and Stern 2002).

The range of response options and strategic interventions that should be applied to a particular problem will depend on such factors as its nature (economic, environmental, or social), its scale (temporal, spatial, or institutional), and the capacity of the actor or decision-maker to make change.

The MA will evaluate the use and effectiveness of various response options and strategic interventions within this context. For example, as knowledge and understanding of the value of ecosystem services increases, the merits of investments to improve or restore ecosystems may become apparent. But in order to attract the financial, human, and social capital—whether public or private—needed to pursue such opportunities, incentives may be required that include the assignment of "property rights" in ecosystem services.

In addition, mainstream economics suggests that a set of property rights that is comprehensive, exclusive, enforceable, and transferable is necessary for efficient outcomes. Yet many of the problems in the economics of the environment can be understood in terms of the failure of systems of property rights to meet this ideal. The failures range from the overexploitation of open access resources to the creation of nuisance or enjoyment to others without compensation (called externalities in the language of economics). Efficient economic outcomes also require perfect information, so that all participants have the same complete information, including about the consequences of their actions. Less-than-perfect information about the functioning of ecosystems can be a significant obstacle to effective choices.

Although there is a tendency to think of property rights in terms of private property, many institutional arrangements in fact create property rights that are not fully private. The sort of village-level institutions that many societies have developed to allocate the rate and intensity of use of common property such as pastureland is an example of this. It is the breakdown or failure of these institutions to evolve that can lead to exploitation of the commons.

There is growing understanding of the functioning of common property resource regimes (Ostrom et al. 2002). Points to consider about the community using the resource include its size and cultural homogeneity, the options for mobility in and out of the community, the frequency of communication between individuals, the density of social networks, practices of reciprocity, and the degree of adherence to shared norms. The characteristics of the resource itself must also be looked at, such as its mobility, its capacity to be stored, and the clarity of its boundaries. These considerations influence the ease and cost of monitoring resource users' behavior and the state of the resource. The continuing state of the common property is the result of all these factors and the ease and cost of enforcing rules about its use.

Gaining access to economic rents from resources in the absence of effective allocation of property rights can be a powerful motivator for many individuals and groups. In many cases, central governments—as the owners of natural resources on behalf of a nation—attempt to monopolize these rents. This may make perfect sense for highly concentrated resources like minerals or crude oil, but for dispersed resources such as forests, control by the central government may stop communities from using local resources. Not surprisingly, people with no property right in local forest resources see little benefit in managing these resources in a sustainable manner.

The pursuit of resource rents helps to explain much of the political economy of the use of ecosystems. Because returns on investment are high when external costs are disregarded, the powerful and those with access to capital have strong incentives to seek these rents. The exercise of political power by individuals, families, and groups in pursuit of resource rents leads to many of the inequities observed in access to and use of natural resources. In addition, where the powerful have the means to exploit natural resources but no legitimate property right, many of the problems of sustainability, of overriding cultural and intrinsic values, and even of efficiency (in terms of broader social welfare) can be explained. Response options and strategic interventions that align property rights in ways that consider all stakeholders or that internalize all costs will be a focus of the MA.

Response options and strategic interventions can be implemented through a number of mechanisms, including international conventions; multilateral and bilateral treaties; national and local laws, regulations, and enforcement; institutional change and changes in governance structures; governmental and industrial policies; contractual agreements, partnerships, and collaboration; and private and public action.

International agreements concerned with ecosystems range from general principles, such as those contained in global framework conventions, to detailed regulatory arrangements with compliance provisions. When negotiated and approved by sovereign states, in principle these agreements constitute the boundary conditions for all related prevailing social, economic, and political national policies. In many cases, however, these conditions will depend on enactment and enforcement of laws and regulations that are designed to implement a nation's responsibilities under the agreement. A literature is emerging on the implementation and effectiveness of such treaties and conventions (Victor et al. 1998).

National-level decision-making has a special role in several respects. First, even the best-designed local or regional actions are likely to be ineffective in the absence of proper coordination (for example, a stringent and enforced protective measure in one region may simply shift a harmful activity to another region). Second, key legislative power often is anchored at the national level (although the distribution between the national and sub-national levels varies among nations). Finally, nations are the recognized parties in the increasing number of international negotiations and agreements. Nevertheless, they face domestic constraints with respect to policy-making because of the ability of sub-national entities—regional or sectoral, and government or nongovernmental—to influence processes and outcomes. Government policies to protect ecosystems can fail if they are at odds with the prevailing social reality: poverty-stricken communities have little to lose by ignoring laws on protected areas if no alternative source of living is provided and if enforcement is weak.

There also are many policies emerging among networks of private-sector firms that may have substantial impact on ecosystems and their services (Dietz and Stern 2002). These include a variety of agreements that set standards and codes of practice for the extraction of resources and the production of goods. Such policies are sometimes applied within a single nation, but there are an increasing number of international agreements as well. They are voluntary but are often coordinated with governments, international agencies, and environmental nongovernmental organizations. Just as with national and international policies, the private-sector agreements may be undermined by local economic circumstances, by a lack of technical capacity at the local level, or by conflicting performance standards within and between private firms.

One important type of strategic intervention that requires assessment is incorporation of the value of ecosystem services into decision-making. Decisions and actions that have direct or indirect effects on ecosystems are usually taken with human well-being in mind, but it is not certain that human well-being (taken broadly) has been enhanced if ecosystem effects have not been taken into consideration. For example, it is useful to think of two kinds of human actions that affect ecosystems and their services: local action that changes ecosystem services directly, and the actions of a number of individuals across a locale, region, or the planet, which produce effects that can be cumulative, dispersed, indirect, but in fact systematic. Humans change biological and chemical cycles, disperse both synthetic and natural chemicals in new ways, and alter planetary processes such as climate and the incidence of ultraviolet radiation. These cumulative or indirect effects are often unintentional, but they can have substantial impacts on ecosystems throughout the globe.

The bulk of the ecosystem degradation literature is concerned with direct interventions, but it is increasingly important to consider how indirect actions influence ecosystems, how these changes interact with direct changes, and how to develop policies and management strategies to deal with each. Conversely, many decisions that influence ecosystems made at the local level are shaped by regional, national, and global conditions (Vayda 1988; Dietz and Rosa 2002). When an ax swings to cut a tree in a forest, there is a sense in which the decision to clear land is made locally. But that decision is influenced by physical infrastructure (such as roads and mills) and by institutions (markets, enforcement of property rights, and land protection regimes) that are in turn shaped by regional, national, and global circumstances. So while decision-making is local, the local action can be shaped by global forces and have global implications.

The extent to which a full range of costs and benefits, including ecosystem services and effects, both direct and indirect, are incorporated in decision-making processes—that is, are felt by the decision-maker—determines the quality of those processes. Therefore, the efficacy and need for strategic interventions that aggregate and focus these costs and benefits on the "local decision-maker" should be assessed. In this regard, as the full costs of action have been concentrated in this way, markets will pass them to ultimate consumers to help them become informed about ecosystem effects and influenced in their consumption behavior. Markets for carbon emissions and sequestration credits are an interesting example of capturing costs and benefits that are otherwise external and making them available to local decision-makers.

Usable Knowledge

A simplified picture of the role of knowledge in decision-making is shown in Figure 8.1, which portrays three interacting processes: monitoring, the decision-making cycle, and the flow of information to and from stakeholders. Policy-making starts by identifying a problem, then it defines policy options and their choice, formulation, and implementation, and ideally it finishes with monitoring and evaluation of the results of executed actions. The process is interactive and iterative and takes place within a specific institutional structure. At all stages, decisions are based on the values, preferences, intuitions, prejudices, and social situations of the organizations and individuals who make them. The process engages all "stakehold-

FIGURE 8.1   Information in the Decision-making Cycle
See text for explanation.

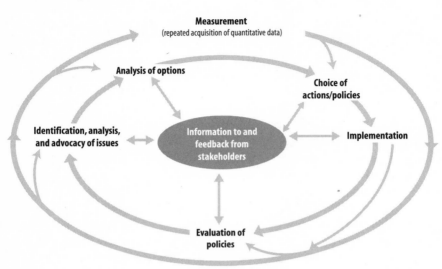

ers," including effective delivery of essential information to decision-makers, communication among stakeholders, and multidirectional exchanges among information providers and information users.

Measurement assembles information from regular monitoring (the outer cycle in Figure 8.1) and other sources. The identification, analysis, and advocacy of issues all require comprehensive and detailed knowledge of human (socioeconomic) and environmental conditions and major trends, including the nature, distribution, and impact of direct and indirect drivers. Hence they need to draw on accounts, spatial assessments, a comprehensive indicator-based assessment, and sometimes also a science assessment. (See Box 8.1.)

The same tools are required for the analysis of options and the choice of actions or policies. They provide the detailed knowledge necessary to examine which issues to address and in what ways, taking account of feasibility, cost-effectiveness, and the likely impacts of different options on socioeconomic and environmental conditions as well as on particular stakeholders.

Policies are implemented through institutions. An institutional analysis is necessary to identify the constraints on implementation and what needs to be done to overcome them. Because implementation depends heavily on the active support and participation of stakeholders, they need

BOX 8.1   Accounts and Assessments

Accounts are assemblages of numerical data, converted to a common unit (money, weight, area, or energy). They can produce valuable composite indicators (constructed directly from data), such as the gross domestic product, genuine saving, and the ecological footprint.

Spatial assessments are assemblages of spatial data. They use geographic information systems to show the location, size, pattern, condition, and ecological, economic, or cultural values and characteristics of land and water areas. They provide basic information for the allocation of uses and are a means of compiling useful composite indicators, such as the status of ecosystem diversity, the extent and security of ecosystem protection, and the extent and severity of land degradation.

Indicator-based assessments are assemblages of indicator variables. Because they rely on representative indicators, they can be selective, and thus they can cover the wide array of issues necessary for an adequate portrayal of human well-being, environmental conditions, and human-environment interactions. Indicators of success are derived:

- in the biophysical/ecological sciences from different kinds of environmental data,
- in sociology and anthropology from concepts of social stability or resilience among individuals or higher organizational units,
- in political sciences from the efficiency with which policies are implemented,
- in jurisprudence from the extent of compliance with law, and
- in economics from the impacts of policy on social welfare.

Because of the confusing and often conflicting signals sent by a large number of individual indicators, assessment methods that produce indexes (compound indicators or combinations of lower-level indicators) are much easier to interpret and can provide decision-makers with clearer and more compelling information. Examples are the Human Development Index prepared annually by the United Nations Development Programme and the Wellbeing Index put together by the International Development Research Centre of Canada and others.

Science assessments use a mixture of numerical data, spatial data, and indicator variables to formulate a scientific consensus on major issues. Whereas other evaluations are conducted regularly, science assessments tend to be produced occasionally, as the need arises. A recent example is the reports of the Intergovernmental Panel on Climate Change. Similarly, European Environment Agency reports collect, analyze, and report data on the state and direction of environmental quality in the entire European region. Models and integrated assessments conducted for the Convention on Long-Range Transboundary Air Pollution give feedback to negotiators on policy options. They can have many different impacts on the policy process: change the terms of a debate (by introducing new policy options, for example), prompt new participants to be concerned about an issue, or change the interests, behavior, or strategies of current participants.

to be informed and feedback should be obtained from them at every stage in the decision-making cycle.

Monitoring and indicator-based assessments track implementation, recording:

- whether actions or policies were implemented,
- whether they achieved their intended results, and
- whether new factors have arisen, in which case the entire cycle is repeated.

Failure to implement requires examining whether the policy was correct, the necessary constituency developed, the instruments put in place, and—if all that happened—the instruments were appropriate. If the relevant indicators used by the indicator-based assessment are unlikely to change in time, one or more intermediate or proximate indicators will be needed to establish a causal link between the actions or policies and the intended results in terms of their impacts on ecosystems and human well-being. This may be complex, as changes in the state of ecosystems and provisioning of services can be caused by several factors operating simultaneously, such as parallel policies, or by external factors such as changes in economic activities. Also, ecosystems are dynamic by nature, and human-induced changes cannot always be distinguished from natural ones. Time lags between responses and ecosystem improvement or change can be considerable, and therefore it is important to evaluate impacts on direct and indirect drivers as well.

Analytical frameworks, such as that developed by the European Environmental Agency (EEA), can be built upon in a MA-type assessment of response options (EEA 2001). The EEA framework distinguishes between the various components of policy development, implementation of measures, strategies, interventions, and ultimate impacts on ecosystems and society. It also indicates some of the key issues or questions that need to be addressed. The design and structure of objectives affect the resource requirements (financial and human), which in turn will affect the efficiency of policy outputs. In this framework, the needs of society, the impacts on the environment, and the outcomes of policies are external to the policy development process. The evaluation of different responses is always in terms of relevance of objectives and the ultimate welfare of society as a result of implementation of these responses. (Issues related to evaluating the policy-making process are discussed in the first section of this chapter.)

To be usable, knowledge needs to address the particular concerns of a user. In the context of the MA, information should have a clear connection to direct and indirect drivers, ecosystem services, and human well-being. General characteristics of good indicators are described in Chapter 7; other characteristics of indicators useful for policy-making are that they:

- relate directly to policy options, goals, or targets (such as the Millennium Development Goals);
- capture change over time;
- identify critical thresholds or the irreversibility of a change;
- provide early warning; or
- characterize the optimal, sufficient, or insufficient level of a given ecosystem service.

It is important to keep in mind that usable knowledge deals with different spatial scales, time frames, and organizational levels. The principal findings are seldom easily transferable from one scale or level to another. Indeed, in most cases the transfer of information across scales needs a special effort. One example is an evaluation of the regional or local impact of global climate change or other global phenomena. The recently stressed notion of a "place-based" science for sustainability—which should be relevant for local policy-making—points in this direction (ICSU 2002a). It is equally important—and difficult—to translate long-term impacts that may affect only future generations into terms relevant to day-to-day decision-making.

The MA will produce a wealth of policy-relevant, preferably quantitative data. This does not mean that everything must be quantified. Indeed, as noted earlier, some elements of sustainable development are intrinsically hard to quantify, and not everything can be turned into numerical data or graphical expressions. This is also true of information and knowledge on some social and economic assets. It is impossible to express in a credible way and in quantitative terms the intrinsic value of biodiversity or the nature of social relationships. But to avoid neglecting them, it is necessary to provide qualitative ways of gathering and communicating information, such as ethnographies of collective actors, assessment of cultural dimensions, case studies, qualitative studies of corruption, and qualitative surveys.

But the majority of "usable knowledge" is in the form of numerical or other quantitative information (ICSU 2002b). Among various forms of such information, indicators play an important role. Indicators provide the basis for assessing progress toward sustainable development. Long-term targets only have meaning as policy goals if progress toward them can be assessed objectively. This requires targets expressed in precise terms. Careful measurement will also improve the ability to identify interactions between different policies and deal with possible trade-offs. These are already a part of policy-making, but the advantage of measurement is that trade-offs are made explicit and transparent.

The MA aims to incorporate both formal scientific information and traditional or local knowledge. Traditional societies have nurtured and refined systems of knowledge of direct value to those societies but also of considerable value to assessments undertaken at regional and global scales. This information often is unknown to science, and can be an expression of other relationships between society and nature in general and of sustainable ways of managing natural resources in particular (ICSU 2002c). To be credible and useful to decision-makers, all sources of information, whether scientific or traditional, must be critically assessed and validated as part of the assessment process through procedures relevant to the form of knowledge.

The findings of an assessment are likely to be used if they are acceptable to potential users or if the users at least regard the sources and process to be legitimate. A legitimate source is one judged to be so by the knowledge system concerned, whether a scientific discipline, a government, or a tradition. (Science has a way of establishing the legitimacy of its knowledge, and traditional societies have ways of establishing the legitimacy of the knowledge within a particular culture. But methods that apply across cultures or to science and traditional knowledge together do not yet exist.) A legitimate assessment process is one that satisfies users that it is fair and that their interests have been taken into account. So-called global assessments may be questioned by less powerful countries, for instance, because they feel their input was not included or that their interests were ignored; this corresponds to a lack of legitimacy (EEA 2001). This applies also to information of other kinds at national or local levels.

In some cases, wide gaps may exist between the sources of usable knowledge and the potential users. Organizations that synthesize and translate scientific research and explore its policy implications are able to bridge this gap. They are sometimes called "boundary organizations" because they facilitate the transfer of usable knowledge between science and policy and they give both policy-makers and scientists the opportunity to cross the boundary between their domains. Journalists have a similar bridging responsibility to ensure that science and policy information is transmitted to the public in ways that are both objective and engaging. Capacity building is desirable in both these areas.

Dealing with Risk and Uncertainty

Risk refers to the probability that certain actions or decisions will result in harm to humans or have adverse effects on their well-being. Since risk is inherent in all human activity and is usually associated with efforts to secure greater benefits for human well-being, it cannot be eliminated from human progress, technology development, or social innovation. But the assessment of risk, including ecological risk assessment, now has an advanced set of tools for comparing apparently dissimilar environmental threats, options for balancing risks and benefits and the potential trade-offs, and means for ensuring equitable management policies or actions aimed at enhancing the situations of the poor and other vulnerable groups (Jaeger et al. 2001; Dietz et al. 2002a). Risk assessment has significant potential for informing the decision process, particularly when decisions are highly complex and uncertain.

The assessment process has several functions. The first is analysis, to provide the knowledge base needed to support sound decisions. This should draw, as noted throughout this report, on both scientific and traditional or lay sources of knowledge to identify and characterize the benefits and risks that various human actions or decisions will have for ecosystem services and human well-being. It should also identify alternative decision options aimed at enlarging benefits, minimizing or eliminating risks, or securing greater fairness in the distribution of benefits and risks. Such analysis should include specific appraisal of the types and magnitudes of uncertainty associated with the estimates.

A second function of assessment is that of deliberation, which is an important attribute of the process (Stern and Fineberg 1996; Dietz and Stern 1998). Deliberation refers to the consultative process and stakeholder involvement, which helps ensure completeness and inclusiveness in the values that different people attach to potential benefits and risks.

Many decisions involved in the management of ecosystems involve high levels of uncertainty or even ignorance. This has led to increased interest in a certain strategy of decision process, which can be described as adaptive management. This approach begins with the recognition that the decision situation or the management challenge is only partly knowledge, and that high levels of uncertainty or ignorance will continue to characterize the situation. In such cases, there are many advantages to structuring the decision process as on ongoing set of interventions that are essentially experiments, as described earlier, and then learning about the relationships based on the outcomes of the decision. This assumes that surprises and unexpected events occur and that management needs to be highly responsive and flexible rather than attempting to control and eliminate variability and uncertainty. The principles of adaptive management and the relevant experience in the Columbia River Basin have been examined by Lee (1993) and elaborated by others (Gunderson et al. 1995b).

Concern that the large uncertainties accompanying the threats to ecosystems and related human well-being will lead to long delays in decision-making and management response has led to increased use of the precautionary principle. As defined in Principle 15 of the 1992 Rio Declaration, this means that "[w]here there are threats of serious or irreversible damage, lack of full scientific certainty shall not be used as a reason for postponing cost-effective measures to prevent environmental degradation" (United Nations 1992:3). A variety of versions of this principle are now in use, with differing implications for assessment and decision-making. Though some commentators have viewed the precautionary principle as an alternative to risk analysis, it is in effect an ethical principle for particular decision situations that is largely compatible with risk analysis. Risk assessment provides valuable knowledge for when to invoke the precautionary principle and the form it might take, but this should not preclude the continuing development of risk-related knowledge to be used in future decisions.

Risk assessment and risk management techniques are often used in the broader processes of environmental impact assessment and strategic environmental assessment. The former is the process of evaluating possible environmental impacts of a proposed project, covering all possible harmful and favorable socioeconomic, cultural, and health-related impacts. Most countries have legislation requiring an environmental impact statement before a project or development is authorized, but the enforcement, practice, and quality requirements for the process vary widely across countries and even across regions within a country. Strategic environmental assessments identify and evaluate the possible environmental implications of proposed policies, broader programs, or large-scale plans in a comprehensive and systematic manner. Their scope ranges from overall sectoral policies (such as a national water policy) to comprehensive regional development strategies. They often provide the context and the background information for project-specific environmental impact assessments.

At a more comprehensive level, communities, nations, groups of nations, or international organizations regularly produce State of the Environment reports to assess environmental trends and conditions and the performance of existing environmental regulations as well as to help formulate new or revised environmental targets and policies. Such reports often identify newly emerging issues or dangerous trends that would be investigated in a strategic environmental assessment in more detail. Accordingly, the three types of activities are closely related and represent key features of the decision-making processes dealing with the interactions between people and ecosystems.

Decision Analytical Frameworks and Tools

The diverse characteristics of the decision-making situations associated with ecosystem and biodiversity management imply the need for a range of decision analytical frameworks (DAFs) and tools. A decision analytical framework is defined as a coherent set of concepts and procedures aimed at synthesizing available information from relevant segments of an ecosystem management problem in order to help policy-makers assess consequences of various decision options. DAFs organize the relevant information in a suitable framework, apply a decision criterion (based on some paradigms or theories), and identify the best options under the assumptions characterizing the analytical framework and the application at hand. It is important to note that none of the frameworks can incorporate the full complexity of decision-making; hence their results supply only part of the information shaping the outcome. And there are always hidden value judgments involved in the selection and application of DAFs.

A broad range of frameworks can be used in principle and has been used in practice to provide information for policy-makers concerned with ecosystem-related decisions at various levels. Based on Toth (2000), Table 8.1 provides an exemplary rather than an all-encompassing list. (See the MA Methods report for concise descriptions of these frameworks.) Many DAFs overlap in practice. Further, one method of analysis often requires input from other methods. As a result, a clear classification of methods and their application to real-world problems is sometimes difficult.

DAFs can be divided into several types: normative DAFs, such as decision analysis and cost-benefit analysis, that deal more directly with valuation and commensuration; descriptive DAFs that consider outcomes that may result from certain actions, such as game theory; and deliberative DAFs that deal with the discovery of information from people and by

TABLE 8.1   Decision Analytical Frameworks

| Framework | Decision Principles | | | Treatment of Uncertainty | | Level of Application | Domain of Application |
	Optimization/ Efficiency	Precautionary Principle	Equity	Rigor	Form		
Decision analysis	*	+	+	*	St	X	B
Cost-benefit analysis	*	−	+	+	SA	X	D
				*	Sc		
Cost-effectiveness analysis	*	+	+	+	SA	X	D
				*	Sc		
Portfolio theory	*	+	−	*	St	X	D
Game theory	+	−	+	+	SA	X	I
				*	St		
Public finance theory	*	−	*	−	SA	N-R	D
Behavioral decision theory	−	+	+	−	Sc	N-M	B
Policy exercises	+	+	+	+	Sc	X	B
Focus groups	−	+	+	−	Sc	R-M	B
Simulation-gaming	−	+	+	+	Sc	X	B
Ethical and cultural prescriptive rules	−	+	+	−	Sc	N-M	D

Compatibility with/usability of decision principles in DAFs:
− weak but not impossible   + possible but not central   * essential feature of DAF

Level of application:
G = Global   I=Inter/Supra-national   N=National   R = Regional/Sectoral (Sub-national)   L=Local (community)
M = Micro (Family, firm, farm)   X = All

Typical domain of application:
D=Direct intervention   I=Indirect influence   B=Both

Uncertainty treatment:
Rigor: * high   + good   − moderate/low
Form: St=Model structure   SA=Sensitivity analysis   Sc=Scenarios

people, such as simulation-gaming. A number of DAFs, such as behavioral decision theory or portfolio theory, have elements that may be described as either normative or descriptive. Finally, there are the DAFs in traditional and transitional societies that can be typified as ethical and cultural.

Several factors determine what type of DAF can be applied and what sort of framework can provide useful information for decision-making. The context of the decision incorporates social, economic, and environmental dimensions. Most of the decisions affecting ecosystems are private ones made by individuals (as owners, operators, or users) or by firms focusing on efficiency and attempting to maximize expected returns. Such decisions are heavily influenced by the prevailing social norms and aspirations, however, and by existing rules and institutions.

An important part of the context in which private decisions take place is the existing set of rules and regulations put in place by public policies. Modern (as opposed to traditional) societies have established procedures to assess the environmental, social, and economic implications of different public decision options. They also tend to have legally prescribed or routinely adopted decision analytical frameworks to choose among the options according to widely accepted criteria for public policy. But these procedures are usually restricted to decisions of an immediate nature (such as building concessions or emission rights). Impacts from diffuse sources and cumulative impacts such as excessive land depletion are generally dealt with less efficiently. In contrast, many societies in transition economies and in developing countries do not have such established procedures; ecosystem decisions therefore appear to be more arbitrary. In many countries, both industrial and developing, short-sighted or outright flawed public policies often lead to private actions with disastrous consequences for ecosystems. Complex management situations and severe ecosystem disruptions arise from the clashes of traditional and modern societies and during the transition from the former to the latter.

The criteria considered important in any decision situation form different decision-making principles. The predominant criteria for a socially desirable or at least widely accepted decision outcome are rooted deeply in the historical traditions of managing the given ecosystem, in the prevailing social conditions (ranging from the values local actors attach to ecosystem services to the existence and enforceability of property rights and government regulations), and in the economic conditions (level of development, distribution of income, and access to resources and social services). These factors need to be considered carefully when determining the decision-making principles to guide the choice of a decision analytical framework. These principles can be used individually or in combinations as DAFs are adopted to address specific ecosystem problems. Table 8.1 indicates some general decision-making principles and their compatibility with relevant DAFs. It is clear that some DAFs can accommodate some decision principles better than others, but full incompatibility is rare.

Key characteristics of ecosystem decision problems are the spatial and temporal scales involved. They determine the jurisdictional level at which the frameworks appear to be most helpful. Table 8.1 also contains entries regarding the decision-making level at which the given DAF can be applied.

There are many ways to affect individual and social behavior determining ecosystem management. Some decision analytical frameworks are more suitable in support of decisions to regulate management directly; others are more helpful in sorting out decisions that will affect broader behavioral choices. Accordingly, another series of entries in Table 8.1 indicates whether the DAF at hand is applicable to decisions concerned with ecosystem management directly or for broader policies that influence primary or proximate drivers.

Additional key features of the ecosystem decision problem are related to the level of complexity and uncertainty involved and the availability of data. Two columns in Table 8.1 provide indications of the ability of the framework to address uncertainties. The first series indicates the level of rigor (high, good, or moderate/low) at which the given framework can treat uncertainties. The second class of entries shows the typical form adopted for uncertainty analysis in the framework (inherent in the model structure, as in classic decision analysis; parametric or Monte Carlo-based sensitivity analysis; or scenarios).

There is no formal assessment or "decision analysis" in traditional societies. In some circumstances, environmental, demographic, economic, and technological forces lead to unsustainable practices in traditional societies (e.g., Krech III 1999). But many indigenous peoples have been managing their ecosystems in a sustainable manner for centuries or even millennia (Ostrom et al. 2002). The information basis of their management practices was grounded in long-standing experience, conscious observation, and inadvertent "experiments" triggered by natural events or human incidence. The guidelines distilled from these very long-term observations were incorporated into religious rules, cultural rituals, and other social-behavioral principles. From time to time, collisions between traditional societies and ecosystems led to ecosystem degradation (mostly temporary, sometimes permanent) or to social disruptions that were resolved by changing management practices, technologies, or social arrangements.

Rapid socioeconomic changes overwhelming the institutional capacities of ecosystem management have caused the largest shifts in ecosystem structure, function, diversity, and productivity. This situation characterized the period when modern societies first encountered previously unknown regions. The inadvertent introduction of alien species, ranging from microscopic pathogens (against which local people and ecosystems were not immune) to mammals (for example, rodents abundant on ships), and the deliberate introduction of new value systems (precious metals), management practices (cut and run), and technologies (chain saw) disrupted traditionally established balances between society and ecosystems. Similar processes can still be observed in the "transition" societies of many developing countries: the old values and rule systems concerning ecosystem management have broken down but no new rules or enforcement capacities have been put in place. These societies and social situations clearly require assessment and decision analytical capacities in order to establish the new ecosystem management rules and organize their enforcement, but there is hardly any sign of such efforts due to the lack of resources and, often, interest.

Only recently has the need to integrate indigenous ecological knowledge into ecosystem assessments and into developing resource management plans been recognized (e.g., Agrawal 1995; Appiah-Opoku and Mulamoottil 1997; Hellier et al. 1999). Often this recognition comes late. Actual efforts undertaken are characterized by varying degrees of integrity and intensity. The increasing assimilation of indigenous peoples, even in remote rural regions, into the modern socioeconomic system has greatly eroded traditional ecological knowledge—in many regions, irreversibly. If indigenous institutions of ecosystem management (social, political, and judicial institutions and religious beliefs, norms, and practices) have also largely disappeared, it does not make much sense to attempt to reincarnate them. A more sensible strategy is likely to be to firmly establish modern institutions and regulatory mechanisms in order to prevent further degradation and possibly to promote restoration of ecosystem quality and services. Given the lack of modern monitoring equipment and scientific data about ecosystems, traditional ecological knowledge may well make valuable contributions to the development of modern management strategies in these areas.

In contrast, in regions where indigenous institutions and knowledge are still reasonably intact and play a significant role in using ecosystem services, it is worth considering how to incorporate them into the modern institutional and regulatory framework. Taboos rooted in religion, harvesting rules overseen by the community, and penalties imposed by the indigenous judicial system are likely to be more effective ways to protect and use ecosystems sustainably than reliance on disrespected, ill-enforced, or corruption-plagued government regulation.

Yet in the dynamics of the real world, as social change and economic transformation are proceeding inexorably, the co-management of ecosystems by traditional rules and modern regulation faces new challenges from time to time. Downs (2000) argues that cultural acceptability of any alternative practice is key, especially at the local scale of the village or indigenous population, so including them in the decision-making is important to achieve sustainability. The selection of DAFs to help craft socially just, acceptable, environmentally effective, and economically efficient policies becomes a particularly delicate task. It is all the more complex because indigenous perceptions and management of ecosystems are far from homogenous. Atran et al. (2002) observe that three groups who live in the same rain forest in Guatemala show profoundly different behaviors, cognition, and social relations in relation to the forest.

In some regions where indigenous communities have persisted on the periphery of modern societies (in the United States, Canada, Australia, and Mexico), there has been increasing concern in recent years about integrating traditional values and knowledge in modern assessment and decision analytical frameworks (Goma et al. 2001; Paci et al. 2002). The ultimate objective is to step beyond assessment and, by acknowledging the rights and incorporating the environmental knowledge of indigenous communities, to make progress towards co-management of ecosystems (Faust and Smardon 2001).

In summary, the choice of the decision analytical framework to support the formulation of policies and measures regarding ecosystem management is influenced by a large number of factors. They range from the social, economic, and cultural context to the geographical and related jurisdictional scale, and from the complexity and uncertainty characteristics of the problem to the preferred nature of the intervention. Advanced analytical frameworks (such as multicriteria decision analysis or cost-benefit analysis) have been widely and successfully used to select among policy options in public and private sectors in many industrial countries. In contrast, some regions with severe environmental problems and high risks of ecosystem degradation are ones in which traditional ecological knowledge and management schemes have faded away but new assessment processes and management systems are still weak or hardly established. In many countries, increasing attempts to combine modern analytical techniques with traditional ecological knowledge, where it still exists, indicate a promising future direction.

APPENDIX 1

Authors

All of the authors of this report have contributed to all chapters of the document. The following list identifies their primary chapter contributions.

Joseph Alcamo, an environmental scientist, is a professor in the Department of Environmental Systems Engineering and Science and Director of the Center for Environmental Systems Research at the University of Kassel in Witzenhausen, Germany. (Ch. 7)

Neville J. Ash, a conservation scientist, is Coordinator for the Millennium Ecosystem Assessment Condition and Trends Working Group at the United Nations Environment Programme World Conservation Monitoring Centre (UNEP-WCMC) in Cambridge, UK. (Ch. 7)

Elena M. Bennett, an ecologist, is a Research Associate at the Center for Limnology at the University of Wisconsin in Madison, Wisconsin, USA. (Ch. 7, contributing author)

Reinette (Oonsie) Biggs, an environmental science masters student at the University of the Witwatersrand in Johannesburg, works as a Research Assistant in the Division of Water, Environment and Forestry Technology at the Council for Scientific and Industrial Research (CSIR) in Pretoria, South Africa. (Ch. 5, contributing author)

Colin D. Butler, an epidemiologist, is a Research Fellow at the National Centre for Epidemiology and Population Health at the Australian National University in Canberra. (Ch. 3)

J. Baird Callicott, an environmental philosopher, is a professor in the Department of Philosophy and Religion Studies at the Institute of Applied Sciences of the University of North Texas in Denton, Texas, USA. (Ch. 6)

Doris Capistrano, a resource economist, is Director of Forests and Governance at the Center for International Forestry Research in Bogor, Indonesia. (Ch. 1, Ch. 5)

Stephen R. Carpenter, an ecologist, is a professor at the Center for Limnology of the University of Wisconsin in Madison, Wisconsin, USA. (Ch. 1, Ch. 2, Ch. 7)

Juan Carlos Castilla, a biologist, is a professor at the Center for Advanced Studies in Ecology and Biodiversity in the Faculty of Biological Sciences of the Pontificia Universidad Catolica de Chile in Santiago. (Ch. 2)

Robert Chambers, a social scientist, is a Research Associate of the Institute of Development Studies at the University of Sussex in Brighton,UK. (Ch. 3)

Poh-Sze Choo, a fisheries scientist, is a Science and Policy Specialist at the WorldFish Center in Penang, Malaysia. (Ch. 2, contributing author)

Kanchan Chopra, an economist, is a professor and Head of the Environmental and Resource Economics Unit at the Institute of Economic Growth in Delhi, India. (Ch. 1, Ch. 3)

Angela Cropper, a development specialist, is President of the Cropper Foundation in Trinidad and Tobago. (Ch. 1)

Gretchen C. Daily, an ecologist, is an Associate Professor (Research) in the Department of Biological Sciences and Senior Fellow in the Institute for International Studies at Stanford University in Stanford, California, USA. (Ch. 1)

Partha Dasgupta, an economist, is the Frank Ramsey Professor of Economics at the University of Cambridge and Fellow of St. John's College in the United Kingdom. (Ch. 1, Ch. 3)

Rudolf de Groot, an ecologist, is a Senior Researcher at the Environmental Systems Analysis Group in the Department of Environmental Sciences at the Wageningen University in Wageningen, The Netherlands. (Ch. 2, Ch. 6)

Thomas Dietz, a human ecologist and sociologist, is Director of the Environmental Science and Policy Program at Michigan State University in East Lansing, Michigan, USA. (Ch. 8)

Anantha Kumar Duraiappah, an economist, is Director of Economic Policy and Senior Economist at the International Institute for Sustainable Development (IISD) in Winnipeg, Canada. (Ch. 3)

Jonathan Foley, a climatologist and ecologist, is Director of the Center for Sustainability and the Global Environment at the University of Wisconsin in Madison, Wisconsin, USA. (Ch. 7, contributing author)

Madhav Gadgil, an ecologist, is a professor at the Centre for Ecological Sciences of the Indian Institute of Science in Bangalore, India.

Kirk Hamilton, an economist, is Team Leader for Policy and Economics in the Environment Department of the World Bank in Washington, DC, USA. (Ch. 8)

Rashid Hassan, an environmental economist, is a professor and Director of the Centre for Environmental Economics and Policy in Africa at the University of Pretoria in Pretoria, South Africa. (Ch. 1, Ch. 6)

Pushpam Kumar, an environmental economist, is an Associate Professor at the Institute of Economic Growth of the University of Delhi Enclave in Delhi, India. (Ch. 3, contributing author)

Eric F. Lambin, a geographer, is a professor in the Department of Geography at the University of Louvain in Louvain-la-Neuve, Belgium. (Ch. 4)

Louis Lebel, an ecologist, is Director of the Unit for Social and Environmental Research in the Faculty of Social Sciences at the Chiang Mai University in Chiang Mai, Thailand. (Ch. 5)

Marcus J. Lee, an economist, is Coordinator for the Millennium Ecosystem Assessment Sub-global Working Group at the WorldFish Center in Penang, Malaysia. (Ch. 5, contributing author)

Rik Leemans, an ecologist, is Senior Scientist at the Netherlands Environmental Assessment Agency of the National Institute of Public Health and the Environment (RIVM) in Bilthoven; a professor at Wageningen University in Wageningen, The Netherlands; and is Co-chair of the Millennium Ecosystem Assessment Responses Working Group. (Ch. 1, Ch. 4)

Liu Jiyuan, a geographer, is a professor at the Institute of Geographical Sciences and Natural Resources Research at the Chinese Academy of Sciences in Beijing, China. (Ch. 7)

Jean-Paul Malingreau, a tropical agronomist and remote sensing specialist, is in charge of the Work Programme of the Joint Research Centre of the European Commission in Brussels, Belgium. (Ch. 7)

Robert M. May (Lord May of Oxford, OM AC), an ecologist and President of the Royal Society, is a professor in the Department of Zoology at the University of Oxford in Oxford, UK. (Ch. 1)

Alex F. McCalla, an agricultural economist, is a Professor Emeritus in the Department of Agricultural and Resource Economics, University of California, Davis in Davis, California, USA. (Ch. 4)

Tony (A.J.) McMichael, an epidemiologist, is a professor and Director of the National Centre for Epidemiology and Population Health at the Australian National University in Canberra, Australia. (Ch. 3)

Bedrich Moldan, an environmental science and policy specialist, is a professor and Director of the Charles University Environment Centre in Prague, Czech Republic. (Ch. 8)

Harold A. Mooney, a plant ecologist, is the Paul S. Achilles Professor of Environmental Biology at Stanford University in Stanford, California, USA. (Ch. 1, Ch. 2)

Richard H. Moss, a specialist in public policy, is Director of the Office of the Climate Change Science Program in Washington, DC and Staff Scientist at the Joint Global Change Research Institute of the University of Maryland in College Park, Maryland, USA. (Ch. 7, contributing author)

Shahid Naeem, an ecologist, is a professor in the Department of Biology at the University of Washington in Seattle, Washington, USA. (Ch. 2)

Gerald C. Nelson, an economist, is an Associate Professor in the Department of Agricultural and Consumer Economics at the University of Illinois in Urbana-Champaign, Illinois, USA. (Ch. 4)

Niu Wen-Yuan, an ecologist, is a professor at the Chinese Academy of Sciences in Beijing, China. (Ch. 3)

Ian Noble, an ecologist, is Senior Advisor to the Carbon Finance Unit at the World Bank in Washington, DC, USA. (Ch. 2, contributing author)

Ouyang Zhiyun, an ecologist, is a professor at the Research Center for Eco-Environmental Sciences at the Chinese Academy of Sciences in Beijing, China. (Ch. 2)

Stefano Pagiola, an economist, is Senior Environmental Economist at the Environment Department of the World Bank in Washington, DC, USA. (Ch. 6)

Daniel Pauly, a biologist, is a professor and Principal Investigator of the Sea Around Us Project at the Fisheries Centre of the University of British Columbia in Vancouver, BC, Canada. (Ch. 7)

Steven Percy, a petroleum industry executive (retired), is visiting faculty of the Corporate Environmental Management Program at the University of Michigan in Ann Arbor, Michigan, USA. (Ch. 8)

Gerhard Petschel-Held was trained as a physicist and is now Head of the Department for Integrated Systems Analysis at the Potsdam Institute for Climate Impact Research in Potsdam, Germany. (Ch. 5, contributing author)

Prabhu Pingali, an economist, is Director of Economic Development Analysis Division at the Food and Agriculture Organization of the United Nations in Rome, Italy. (Ch. 1, Ch. 4)

Sarah Porter, a natural resource economist, is a Research Collaborator in the Environment and Production Technology Division at the International Food Policy Research Institute in Washington, DC, USA. (Ch. 7, contributing author)

Robert Prescott-Allen is author of *The Wellbeing of Nations* and Executive Director of the Coast Information Team in British Columbia, Canada. (Ch. 8)

Walter V. Reid, an ecologist and policy analyst, is Director of the Millennium Ecosystem Assessment at the WorldFish Center in Penang, Malaysia. (Ch.1, Ch. 2)

Taylor H. Ricketts, an ecologist, is Director of the Conservation Science Program at the World Wildlife Fund in Washington, DC, USA. (Ch. 7)

Cristian Samper, a biologist, is Director of the National Museum of Natural History at the Smithsonian Institution in Washington, DC and former Deputy Director of the Smithsonian Tropical Research Institute in Panama. (Ch. 1, Ch. 5)

Stephen H. Schneider, a climatologist, is a professor in the Department of Biological Sciences and Co-director of the Center for Environmental Science and Policy at Stanford University in Stanford, California, USA. (Ch. 7, contributing author)

Robert (Bob) Scholes, a systems ecologist, is a Fellow of the Council for Scientific and Industrial Research in South Africa. (Ch. 1, Ch. 5)

Henk Simons, an ecologist, is a scientist at the Netherlands Environmental Assessment Agency of the National Institute of Public Health and the Environment (RIVM) in Bilthoven, The Netherlands and is with the Technical Support Unit of the Millennium Ecosystem Assessment Responses Working Group. (Ch. 4)

Ferenc L. Toth, an economist and policy analyst, is an Associate Professor in the Department of Economic Geography and Natural Resource Economics at the Budapest University of Economic Sciences and Public Administration in Budapest, Hungary and a Senior Research Scholar at the International Institute for Applied Systems Analysis (IIASA) in Laxenburg, Austria. (Ch. 8)

Jane K. Turpie, a resource economist and conservation biologist with an ecology background, is senior lecturer at the Percy FitzPatrick Institute of the University of Cape Town in South Africa. (Ch. 2)

Robert Tony Watson, an atmospheric chemist, is Chief Scientist at the World Bank in Washington, DC, USA. (Ch. 1, Ch. 4)

Thomas J. Wilbanks, a geographer, is a Corporate Research Fellow and Leader of Global Change and Developing Country Programs at the Oak Ridge National Laboratory in Oak Ridge, Tennessee, USA. (Ch. 5)

Meryl Williams, a fisheries scientist, is Director General of the WorldFish Center in Penang, Malaysia. (Ch. 2)

Stanley Wood, an agriculture and natural resource specialist, is a Senior Scientist at the International Food Policy Research Institute in Washington, DC, USA. (Ch. 7)

Zhao Shidong, an ecologist, is a research professor at the Synthesis Research Center of Chinese Ecosystem Research in the Institute of Geographic Science and Natural Resources Research at the Chinese Academy of Sciences in Beijing, China. (Ch. 1, Ch. 2)

Monika B. Zurek, an agricultural biologist and economist, is a Post-doctoral Fellow with the Economics Program of the International Maize and Wheat Improvement Center (CIMMYT) in Mexico City, Mexico and is with the Technical Support Unit of the Millennium Ecosystem Assessment Scenarios Working Group. (Ch. 4, Ch. 7)

APPENDIX 2

Reviewers

By country of residence

Tundi Agardy, United States
Peder Agger, Denmark
Heidi J. Albers, United States
Jacqueline Alder, Canada
Dolors Armenteras, Colombia
Ahmad Badkoubi, Iran
Beril Balantekyn, Turkey *
Jan Barkmann, United States
Ivar Baste, Kenya
Gordana Beltram, Slovenia *
Fikret Berkes, Canada
Gunilla Björklund, Sweden
Victor Brovkin, Germany
David W. Cash, United States
Lena Chan, Singapore *
Roberto Moreira Coimbra, Brazil *
Flavio Comim, United Kingdom
Ulisses Confalonieri, Brazil
Carlos Corvalan, Switzerland
Robert Costanza, United States
Philippe Crabbé, Canada
Chris Crossland, Netherlands
Philippe Cury, South Africa
Brian Davies, Canada
Ruth DeFries, United States
Timothy J. Downs, United States
Ann E. Edwards, United States
Thomas Elmqvist, Sweden
Daniel P. Faith, Australia
Marianne Feldmann, Germany *
Colin Filer, Australia
Max Finlayson, Australia
Simon Foale, Australia
Sandy Gauntlett, New Zealand
Habiba Gitay, United States
Matija Gogala, Slovenia
Ann Hamblin, Australia
Arne Sveinson Haugen, Norway
Lars Hein, Netherlands
Ole Hendrickson, Canada *
Georgi Hiebaum, Bulgaria
Joanna I. House, Germany

Robert Howarth, United States
Nay Htun, United States
Jikun Huang, China
Brian Huntley, South Africa
IUCN, Switzerland
Peter Kareiva, United States
G.B. Kasali, Zambia
Thaya Kulenthran, Malaysia
Rodel D. Lasco, Philippines
Anna Lawrence,
    United Kingdom
Patricia Balvanera Levy, Mexico
Michel Loreau, France
Wolfgang Lutz, Austria
David MacDevette, Kenya
Jens Mackensen, Kenya
Peter J. Marcotullio, Japan
Victor H. Marin, Chile
Pim Martens, Netherlands
Jeffrey A. McNeely, Switzerland
Carmen Miranda, Bolivia
Monirul Qader Mirza, Canada
Lisa Moore, United States
Christian Nelleman, Norway
Valery Neronov,
    Russia Federation
Madiodio Niasse, Burkina Faso
Masahiko Ohsawa, Japan
Cheryl Palm, United States
Kirit S. Parikh, India
Henrique Miguel Pereira,
    Portugal
Charles Perrings, United Kingdom
Ian Perry, Canada
Stuart Pimm, United States
Juri Puzachenko,
    Russia Federation
David Rapport, Canada
Paul Raskin, United States
Ginger Rebstock, United States
Kent Redford, United States
Carmen Revenga, United States

Janet Riley, United Kingdom
Jon Paul Rodríguez, Venezuela
Dana Roth, United States
Dale S. Rothman, Netherlands
Lech Ryszkowski, Poland
Uriel Safriel, Israel
Abilio Rachid Said, Guinea-Bissau
Odd Terje Sandlund, Norway
M. Sanjayan, United States
Michael Scherer-Lorenzen,
    Germany
Ernst-Detlef Schulze, Germany
Kate L. Sebastian, United States
Megumi Seki, Kenya
David R. Simpson, United States
Ashbindu Singh, United States
Tone Solhaug, Norway *
Shiv Someshwar, United States
David Stanners, Denmark
Derek Staples, United Kingdom
Salah Tahoun, Egypt *
Lee M. Talbot, United States
Mohamed Tawfic Ahmed, Egypt
Tonnie Tekelenburg, Netherlands
Bakary Toure, Mali *
Dechen Tshering, Bhutan *
Amy N. Van Buren, United States
Annemarie van der Heijden, *
    Netherlands
Charles Vörösmarty, United States
J. Wiens, United States
Bruce Wilcox, United States
Clive Wilkinson, Australia
Matthew A. Wilson, United States
Zerihun Woldu, Ethiopia
Alistair Woodward, New Zealand
Nicolaos Yassoglou, Greece *
Carey Yeager, Indonesia
Masatoshi Yoshino, Japan
Ebil Yusof, Indonesia
Georgy Zavarzin,
    Russian Federation

---

* Representing national focal point

APPENDIX 3

Abbreviations and Acronyms

CBD	Convention for Biological Diversity
CGE	computable general equilibrium (model)
CV	contingent valuation
DAF	decision analytical framework
DPSIR	driver-pressure-state-impact-response
EEA	European Environment Agency
EGS	ecosystem global scenario
EIA	environmental impact assessment
ESA	Endangered Species Act (of the United States)
FWS	Fish and Wildlife Service (of the United States)
GEO-3	*Global Environmental Outlook 3*
GCM	general circulation model
GSG	Global Scenario Group
IP	International Paper
IPAT	impacts = population × affluence × technology
IPCC	Intergovernmental Panel on Climate Change
ISEH	International Society for Ecosystem Health
IUCN	World Conservation Union
MA	Millennium Ecosystem Assessment
NGO	nongovernmental organization
OECD	Organisation for Economic Co-operation and Development
PSIR	pressure-state-impact-response
SMS	safe minimum standard
SRES	Special Report on Emissions Scenarios (of the IPCC)
TEV	total economic value
UNEP	United Nations Environment Programme
WBCSD	World Business Council on Sustainable Development
WSSD	World Summit on Sustainable Development
WTA	willingness to accept
WTP	willingness to pay
WWV	World Water Vision

APPENDIX 4

Glossary

Adaptive management: The mode of operation in which an intervention (action) is followed by monitoring (learning), with the information then being used in designing and implementing the next intervention (acting again) to steer the system toward a given objective or to modify the objective itself.

Baseline: A set of reference data sets or analyses used for comparative purposes; it can be based on a reference year or a reference set of (standard) conditions.

Bayesian probability: A subjective characterization of probabilities of outcomes arising from a certain decision.

Benefits transfer: Economic valuation approach in which estimates obtained (by whatever method) in one context are used to estimate values in a different context. This approach is widely used because of its ease and low cost, but is risky because values are context-specific and cannot usually be transferred.

Bias: Systematic error in a data set due to approaches and methods and their application in sampling, investigation, measurement, classification, or analysis.

Biodiversity: The variability among living organisms from all sources including terrestrial, marine, and other aquatic ecosystems and the ecological complexes of which they are part; this includes diversity within and among species and diversity within and among ecosystems.

Biomass: The mass of living tissues in either an individual or cumulatively across organisms in a population or ecosystem.

Boundary organizations: Public or private institutions that synthesize and translate scientific research and explore its policy implications to help bridge the gap between science and decision-making.

Capability: The combinations of doings and beings from which people can choose to lead the kind of life they value. Basic capability is the capability to meet a basic need.

Capacity building: A process of strengthening or developing human resources, institutions, or organizations.

Capital value (of an ecosystem): The present value of the stream of future benefits that a ecosystem will generate under a particular management regime. Present values are typically obtained by discounting future benefits and costs; the appropriate rates of discount are often a contested issue, particularly in the context of natural resources.

Change in productivity approach: Economic valuation techniques that value the impact of changes in ecosystems by tracing their impact on the productivity of economic production processes. For example, the impact of deforestation could be valued (in part) by tracing the impact of the resulting changes in hydrological flows on downstream water uses such as hydroelectricity production, irrigated agriculture, and potable water supply.

Characteristic scale: The typical extent or duration over which a process is most significantly or apparently expressed.

Command and control: The policy framework in which environmental (e.g., emission standards for each source and each pollutant) and resource (catch or logging limits for each site or species) management rules are prescribed by the regulator, leaving little flexibility for actors in the implementation.

Common pool resource: A valued natural or human-made resource or facility in which one person's use subtracts from another's use and where it is often necessary but difficult to exclude potential users from the resource. See also *common property resource*.

Common property resource: A good or service shared by a well-defined community. See also *common pool resource*.

Constituents of well-being: The experiential aspects of well-being, such as health, happiness, and freedom to be and do, and, more broadly, basic liberties.

Conservation value: See *existence value*.

Consumptive use: The reduction in the quantity or quality of a good available for other users due to consumption.

Contingent valuation (CV): Economic valuation technique based on the stated preference of respondents regarding how much they would be willing to pay for specified benefits. A detailed description of the good or service involved is provided, along with details about how it will be provided. CV is designed to circumvent the absence of markets by presenting consumers with hypothetical markets in which they have the opportunity to buy the good or service in question. The methodology is controversial, but widely accepted guidelines for its application have been developed.

Core data set: Data sets designated as having wide potential application throughout the Millennium Ecosystem Assessment process. These data sets will be made available to all working groups and scientists within the program, and their common use will maximize consistency among analyses. Examples include land use, land cover, and population data sets.

Cultural landscape: See *landscape*.

Cultural services: The nonmaterial benefits people obtain from ecosystems through spiritual enrichment, cognitive development, reflection, recreation and aesthetic experience, including, for example, knowledge systems, social relations, and aesthetic values.

Decision analytical framework (DAF): A coherent set of concepts and procedures aimed at synthesizing available information from relevant segments of the given ecosystem management problem in order to help policy-makers assess consequences of various decision options. DAFs organize the relevant information in a suitable framework, apply decision criteria (both based on some paradigms or theories), and thus identify options that are better than others under the assumptions characterizing the analytical framework and the application at hand.

Decision-maker: A person whose decisions and actions can influence a condition, process, or issue under consideration.

Decomposition: The ecological process carried out primarily by microbes that leads to a transformation of dead organic matter into inorganic mater; the converse of biological production. For example, the transformation of dead plant material, such as leaf litter and dead wood, into carbon dioxide, nitrogen gas, and ammonium and nitrates.

Determinants of well-being: Inputs into the production of well-being, such as food, clothing, potable water, and access to knowledge and information.

Direct use value: In the total economic value framework, the benefits derived from the goods and services provided by an ecosystem that are used directly by an economic agent. These include consumptive uses (e.g., harvesting goods) and nonconsumptive uses (e.g., enjoyment of scenic beauty). Agents are often physically present in an ecosystem to receive direct use value. Compare *indirect use value.*

Domain (of scale): The combined range of characteristic scales for a given process in both space and time.

Downscaling: The process of converting data or information at a course resolution to a finer resolution.

Driver: Any natural or human-induced factor that directly or indirectly causes a change in an ecosystem.

Driver, direct: A driver that unequivocally influences ecosystem processes and can therefore be identified and measured to differing degrees of accuracy.

Driver, endogenous: A driver whose magnitude can be influenced by the decision-maker. The endogenous or exogenous characteristic of a driver depends on the organizational scale. Some drivers (e.g., prices) are exogenous to a decision-maker at one level (a farmer) but endogenous at other levels (the nation-state).

Driver, exogenous: A driver that cannot be altered by the decision-maker. See also *endogenous driver.*

Driver, indirect: A driver that operates by altering the level or rate of change of one or more direct drivers.

Ecological footprint: The area of productive land and aquatic ecosystems required to produce the resources used and to assimilate the wastes produced by a defined population at a specified material standard of living, wherever on Earth that land may be located.

Ecological security: A condition of ecological safety that ensures access to a sustainable flow of provisioning, regulating, and cultural services needed by local communities to meet their basic capabilities.

Ecosystem: A dynamic complex of plant, animal, and microorganism communities and their nonliving environment interacting as a functional unit.

Ecosystem approach: A strategy for the integrated management of land, water, and living resources that promotes conservation and sustainable use in an equitable way. An ecosystem approach is based on the application of appropriate scientific methodologies focused on levels of biological organization, which encompass the essential structure, processes, functions, and interactions among organisms and their environment. It recognizes that humans, with their cultural diversity, are an integral component of many ecosystems.

Ecosystem assessment: A social process through which the findings of science concerning the causes of ecosystem change, their consequences for human well-being, and management and policy options are brought to bear on the needs of decision-makers.

Ecosystem boundary: The spatial delimitation of an ecosystem, typically based on discontinuities in the distribution of organisms, the biophysical environment (soil types, drainage basins, depth in a water body), and spatial interactions (home ranges, migration patterns, fluxes of matter).

Ecosystem function: An intrinsic ecosystem characteristic related to the set of conditions and processes whereby an ecosystem maintains its integrity (such as primary productivity, food chain, biogeochemical cycles). Ecosystem functions include such processes as decomposition, production, nutrient cycling, and fluxes of nutrients and energy.

Ecosystem health: A measure of the stability and sustainability of ecosystem functioning or ecosystem services that depends on an ecosystem being active and maintaining its organization, autonomy, and resilience over time. Ecosystem health contributes to human well-being through sustainable ecosystem services and conditions for human health.

Ecosystem interactions: Exchanges of materials and energy among ecosystems.

Ecosystem properties: The size, biodiversity, stability, degree of organization, internal exchanges of materials and energy among different pools, and other properties that characterize an ecosystem.

Ecosystem services: The benefits people obtain from ecosystems. These include provisioning services such as food and water; regulating services such as flood and disease control; cultural services such as spiritual, recreational, and cultural benefits; and supporting services such as nutrient cycling that maintain the conditions for life on Earth. The concept "ecosystem goods and services" is synonymous with ecosystem services.

Ecosystem stability: A description of the dynamic properties of an ecosystem. An ecosystem is considered stable if it returns to its original state shortly after a perturbation (resilience), exhibits low temporal variability (constancy), or does not change dramatically in the face of a perturbation (resistance).

Emergent property: A phenomenon that is not evident in the constituent parts of a system but that appears when they interact in the system as a whole.

Equity: Fairness of rights, distribution, and access. Depending on context, this can refer to resources, services, or power.

Existence value: The value that individuals place on knowing that a resource exists, even if they never use that resource (also sometimes known as conservation value or passive use value).

Extent: The length or area over which observations were made or for which an assessment was made or over which a process is expressed.

Externality: A consequence of an action that affects someone other than the agent undertaking that action and for which the agent is neither compensated nor penalized. Externalities can be positive or negative.

Forecast: See *prediction*.

Freedom: The range of options a person has in deciding the kind of life to lead. Freedom is similar to the concept of capability and can be used interchangeably.

Functional redundancy: A characteristic of species within an ecosystem in which certain species contribute in equivalent ways to an ecosystem function such that one species may substitute for another. Note that species that are redundant for one ecosystem function may not be redundant for others.

Geographic information system (GIS): A computerized system organizing data sets through a geographical referencing of all data included in its collections. A GIS allows the spatial display and analysis of information.

Global scale: The geographical realm encompassing all of Earth.

Good social relations: Social cohesion, mutual respect, good gender and family relations, and the ability to help others and provide for children.

Grain (of a phenomenon): A spatial unit that can be considered internally homogenous. Grain (of observation) is the fundamental (that is, the smallest) unit of observation.

Habitat: Area occupied by and supporting living organisms. Also used to mean the environmental attributes required by a particular species or its ecological niche.

Health: Strength, feeling well, and having a good functional capacity. Health, in popular idiom, also connotes an absence of disease. The health of a whole community or population is reflected in measurements of disease incidence and prevalence, age-specific death rates, and life expectancy.

Hedonic price methods: Economic valuation methods that use statistical techniques to break down the price paid for goods and services into the implicit prices for each of their attributes, including environmental attributes such as access to recreation or clean air. Thus the price of a home may be broken down to see how much the buyers were willing to pay for a home in a neighborhood with cleaner air.

Herbivory: The consumption of plants by animals.

Hierarchical systems: Systems that can be analyzed into successive sets of nested subsystems.

Indicator: Information based on measured data used to represent a particular attribute, characteristic, or property of a system.

Indirect use value: The benefits derived from the goods and services provided by an ecosystem that are used indirectly by an economic agent. For example, an agent at some distance from an ecosystem may derive benefits from drinking water that has been purified as it passed through the ecosystem. Compare *direct use value.*

Institutions: The rules that guide how people within societies live, work, and interact with each other. Formal institutions are written or codified rules. Examples of formal institutions would be the constitution, the judiciary laws, the organized market, and property rights. Informal institutions are rules governed by social and behavioral norms of the society, family, or community.

Instrumental: A means to an end.

Interventions: See *responses.*

Intrinsic value: The value of someone or something in and for itself, irrespective of its utility for someone else.

Irreversibility: The quality of being impossible or difficult to return to, or to restore to, a former condition. See also *option value, precautionary principle, resilience,* and *threshold.*

Kantianism: A theory of ethics that ascribes intrinsic value to rational beings and is the philosophical foundations of contemporary human rights and the extended ascription of intrinsic value to a wide spectrum of nonhuman natural entities, including ecosystems.

Land cover: The physical coverage of land, usually expressed in terms of vegetation cover or lack of it. Influenced by but not synonymous with *land use.*

Land use: The human utilization of a piece of land for a certain purpose (such as irrigated agriculture or recreation). Influenced by but not synonymous with *land cover.*

Landscape: An area of land that contains a mosaic of ecosystems, including human-dominated ecosystems. The term cultural landscape is often used when referring to landscapes containing significant human populations.

Length of growing period: For the purposes of the system definitions used in the Millennium Ecosystem Assessment, this is defined for terrestrial ecosystems as the total number of days in a year during which rainfall exceeds one half of potential evapotranspiration.

Level: The discrete levels of social organization, such as individuals, households, communities, and nations. See also *scale.*

Market failure: The inability of a market to bring about the allocation of resources that best satisfies the wants of society. In particular, the overallocation or underallocation of resources to the production of a particular good or service caused by spillovers or informational problems or because markets do not provide desired public goods.

Megadiversity country: One of 17 countries (Australia, Brazil, China, Colombia, Democratic Republic of Congo, Ecuador, India, Indonesia, Madagascar, Malaysia, Mexico, Peru, Philippines, Papua New Guinea, South Africa, United States, and Venezuela) home to the largest fraction of known species in the world.

Metadata: The collection of information related to the type and characteristics of data sets and their location in a data archive.

Open access resource: A good or service over which no property rights are recognized.

Opportunity cost: The benefits forgone by undertaking one activity instead of another.

Option value: The value of preserving the option to use services in the future either by oneself (option value) or by others or heirs (bequest value). Quasi-option value represents the value of avoiding irreversible decisions until new information reveals whether certain ecosystem services have values society is not currently aware of.

Parasitism: The consumption of one individual by another in which the consumer resides on (ectoparasite) or within (endoparasite) the body of its host or victim.

Passive use value: See *existence value*.

Pastoral system: The use of domestic animals as a primary means for obtaining resources from habitats.

Policy failure: A situation in which government policies create inefficiencies in the use of goods and services.

Policy-maker: A person with power to influence or determine policies and practices at an international, national, regional, or local level.

Pollination: The completion of the sexual phase of reproduction in some plants by the transportation of pollen. In the context of ecosystem services, pollination generally refers to animal-assisted pollination, such as that done by bees, rather than wind pollination.

Precautionary principle: The management concept stating that in cases "where there are threats of serious or irreversible damage, lack of full scientific certainty shall not be used as a reason for postponing cost-effective measures to prevent environmental degradation," as defined in the Rio Declaration.

Precision: The ability of a measurement to be consistently reproduced. Also, the degree of accuracy.

Predation: The consumption of animals by other animals.

Prediction (or forecast): The result of an attempt to produce a most likely description or estimate of the actual evolution of a variable or system in the future. See also *projection* and *scenario*.

Primary production: Assimilation (gross) or accumulation (net) of energy and nutrients by green plants and by organisms that use inorganic compounds as food.

Private costs and benefits: Costs and benefits directly felt by individual economic agents or groups as seen from their perspective. (Externalities imposed on others are ignored.) Costs and benefits are valued at the prices actually paid or received by the group, even if these prices are highly distorted. Sometimes termed "financial" costs and benefits. Compare *social costs and benefits*.

Probability distribution: A distribution that shows all the values that a random variable can take and the likelihood that each will occur.

Projection: A potential future evolution of a quantity or set of quantities, often computed with the aid of a model. Projections are distinguished from "predictions" in order to emphasize that projections involve assumptions concerning, for example, future socioeconomic and technological developments that may or may not be realized; they are therefore subject to substantial uncertainty.

Property rights: An institution that gives someone possession rights to use things and to prevent others from using them; includes private, collective, common, public, and state property rights.

Provisioning services: The products obtained from ecosystems, including, for example, genetic resources, food and fiber, and fresh water.

Rangeland: An area where the main land use is related to the support of grazing or browsing mammals, such as cattle, sheep, goats, camels, or antelope.

Regulating services: The benefits obtained from the regulation of ecosystem processes, including, for example, the regulation of climate, water, and some human diseases.

Reporting unit: The spatial or temporal unit at which assessment or analysis findings are reported. In an assessment, these units are chosen to maximize policy relevance or relevance to the public and thus may differ from those upon which the analyses were conducted (e.g., analyses conducted on mapped ecosystems can be reported on administrative units).

Resilience: The capacity of a system to tolerate impacts of drivers without irreversible change in its outputs or structure.

Resolution (of observation): The spatial or temporal separation between observations.

Responses: Human actions, including policies, strategies, and interventions, to address specific issues, needs, opportunities, or problems. In the context of ecosystem management, responses may be of legal, technical, institutional, economic, and behavioral nature and may operate at local or micro, regional, national, or international level and at various time scales.

Risk: The probability or probability distribution of an event or the product of the magnitude of an event and the probability of its occurrence.

Safe minimum standard: A decision analytical framework in which the benefits of ecosystem services are assumed to be incalculable and should be preserved unless the costs of doing so rise to an intolerable level, thus shifting the burden of proof to those who would convert them.

Scale: The physical dimensions, in either space or time, of phenomena or observations.. See also *level.*

Scenario: A plausible and often simplified description of how the future may develop, based on a coherent and internally consistent set of assumptions about key driving forces (e.g., rate of technology change, prices) and relationships. Scenarios are neither predictions nor projections and sometimes may be based on a "narrative storyline." Scenarios may be derived from projections but are often based on additional information from other sources.

Security: Access to resources, safety, and the ability to live in a predictable and controllable environment.

Social costs and benefits: Costs and benefits as seen from the perspective of society as a whole. These differ from private costs and benefits in being more inclusive (all costs and benefits borne by some member of society are taken into account) and in being valued at social opportunity cost rather than market prices, where these differ. Sometimes termed "economic" costs and benefits. Compare *private costs and benefits*.

Spatial resolution: See *resolution*.

Stakeholder: An actor having a stake or interest in a physical resource, ecosystem service, institution, or social system, or someone who is or may be affected by a public policy.

Statistical variation: Variability in data due to error in measurement, error in sampling, or variation in the measured quantity itself.

Strategies: See *responses*.

Supporting services: Ecosystem services that are necessary for the production of all other ecosystem services. Some examples include biomass production, production of atmospheric oxygen, soil formation and retention, nutrient cycling, water cycling, and provisioning of habitat.

Sustainability: A characteristic or state whereby the needs of the present and local population can be met without compromising the ability of future generations or populations in other locations to meet their needs.

Taxa: Nested groups of species that reflect similarity. Familiar taxa are birds (which belong to the class *Aves*) and fig trees (which belong to the genus *Ficus*).

Taxonomy: A system of nested categories (*taxa*) reflecting evolutionary relationships or morphological similarity.

Threshold: A point or level at which new properties emerge in an ecological, economic, or other system, invalidating predictions based on mathematical relationships that apply at lower levels. For example, species diversity of a landscape may decline steadily with increasing habitat degradation to a certain point, then fall sharply after a critical threshold of degradation is reached. Human behavior, especially at group levels, sometimes exhibits threshold effects. Thresholds at which irreversible changes occur are especially of concern to decision-makers.

Time series data: A set of data that expresses a particular variable measured over time.

Total economic value framework: A widely used framework to disaggregate the components of utilitarian value, including *direct* and *indirect use value*, *option value*, quasi-option value and *existence value*.

Travel cost methods: Economic valuation techniques that use observed costs to travel to a destination to derive demand functions for that destination. Developed to value the recreational use of protected areas, they have limited applicability outside this context.

Uncertainty: An expression of the degree to which a future condition (e.g., of an ecosystem) is unknown. Uncertainty can result from lack of information or from disagreement about what is known or even knowable. It may have many types of sources, from quantifiable errors in the data to ambiguously defined terminology or uncertain projections of human behavior.

Upscaling: The process of aggregating or extrapolating information collected at a fine resolution to a courser resolution or greater extent.

Utilitarian: An approach that focuses on the satisfaction of human preferences. In some cases, this is taken further and made the basis of a moral viewpoint. See also *utilitarianism*.

Utilitarianism: A creed that accepts utility or the greatest happiness as the foundation of morals and holds that actions are right in proportion as they tend to promote happiness.

Utility: In economics, the measure of the degree of satisfaction or happiness of a person.

Value: The contribution of an action or object to user-specified goals, objectives, or conditions.

Value systems: Norms and precepts that guide human judgment and action.

Valuation: The process of expressing a value for a particular good or service in a certain context (e.g., of decision-making) usually in terms of something that can be counted, often money, but also through methods and measures from other disciplines (sociology, ecology, and so on).

Well-being: A context- and situation-dependent state, comprising basic material for a good life, freedom and choice, health, good social relations, and security.

Bibliography

Acheson, J.M., 1993: Capturing the commons: Legal and illegal strategies. In: *The Political Economy of Customs and Culture: Informal Solutions to the Common Problem*, T.L. Anderson and R.T. Simmons (eds.), Rowman and Littlefield, Lanham, MD.

Agrawal, A., 1995: Dismantling the divide between indigenous and scientific knowledge. *Development and Change*, **26(3)**, 413–439.

Agrawal, A., 2002: Common resources and institutional stability. In: *The Drama of the Commons*, E. Ostrom, T. Dietz, N. Dolšak, P.C. Stern, S. Stonich, and E.U. Weber (eds.), National Academy Press, Washington, DC, 41–85.

Alcamo, J., 2001: *Scenarios as Tools for International Assessments*. Prospects and Scenarios No. 5, European Environment Agency, Copenhagen.

Alcamo, J., R. Leemans, and G.J.J. Kreileman, 1998: *Global Change Scenarios of the 21st Century*. Results from the IMAGE 2.1 model. Pergamon & Elsevier Science, London.

Alcamo, J., G.J.J. Kreileman, M.S. Krol, and G. Zuidema, 1994: Modeling the global society-biosphere-climate system, Part 1: Model description and testing. *Water, Air, and Soil Pollution*, **76(March)**, 1–35.

Alcamo, J., G.J.J. Kreileman, R. Leemans, and (eds.), 1996: Integrated scenarios of global change: Results from the IMAGE 2 model. *Global Environmental Change (Special Issue)*, **6(4)**, 255–394.

Alcamo, J., P. Döll, T. Henrichs, F. Kaspar, B. Lehner, T. Rösch, and S. Siebert, 2003: WaterGAP: Development and application of a global model for water withdrawals and availability. *Hydrological Sciences*, (in press).

Alkire, S., 2002: Dimensions of human development. *World Development*, **30(2)**, 181–205.

Allen, T.H.F., 1998: The landscape 'level' is dead: Persuading the family to take off the respirator. In: *Ecological Scale: Theory and Applications*, D.L. Peterson and V.T. Parker (eds.), Columbia University Press, New York, NY, 35–54.

Allen, T.H.F. and T.B. Starr, 1982: *Hierarchy: Perspectives for Ecological Complexity*. University of Chicago Press, Chicago, IL.

Alston, L.J., G. Libecap, and B. Meuller, 1997: Violence and the development of property rights to land in the Brazilian Amazon. In: *The Frontiers of the New Institutional Economics*, J.N. Drobak and J.V.C. Nye (eds.), Academic Press, New York, NY.

Annan, K.A., 2000: *We the Peoples: The Role of the United Nations in the 21st Century*. United Nations, New York, NY.

Appiah-Opoku, S. and G. Mulamoottil, 1997: Indigenous institutions and environmental assessment: The case of Ghana. *Environmental Management*, **21(2)**, 159–171.

Asheim, G., 1997: Adjusting green NNP to measure sustainability. *Journal of Economics*, **99(3)**, 355–370.

Atran, S., D. Medin, N. Ross, E. Lynch, V. Vapnarsky, E. Ek' Ucan, J. Coley, C. Timura, and M. Baran, 2002: Folkecology, cultural epidemiology, and the spirit of the commons: A garden experiment in the Maya Lowlands, 1991–2001. *Current Anthropology*, **43(3)**, 421–450.

217

Ayensu, E., D.R. Claasen, M. Collins, A. Dearing, L. Fresco, M. Gadgil, H. Gitay, G. Glaser, C. Juma, J. Krebs, R. Lenton, J. Lubchenco, J.A. McNeely, H.A. Mooney, P. Pinstrup-Andersen, M. Ramos, P. Raven, W.V. Reid, C. Samper, J. Sarukhán, P. Schei, J.G. Tundisi, R.T. Watson, and A.H. Azkri, 2000: International ecosystem assessment. *Science,* **286,** 685–686.

Babinard, J., 2001: A short history of agricultural biotechnology. In: *Genetically Modified Organisms in Agriculture: Economics and Politics,* G.C. Nelson (ed.), Academic Press, San Diego, CA, 271–274.

Balvanera, P., G.C. Daily, P.R. Ehrlich, T.H. Ricketts, S.A. Bailey, S. Kark, C. Kremen, and H. Pereira, 2001: Conserving biodiversity and ecosystem services. *Science,* **291,** 2047.

Barbier, E.B., 2000: Links between economic liberalization and rural resource degradation in the developing regions. *Agricultural Economics,* **23,** 299–310.

Barr, J., 1972: Man and nature: The ecological controversy and the Old Testament. *Bulletin of the John Rylands Library,* **55,** 9–32.

Barrett, C.B., E.B. Barbier, and T. Reardon, 2001: Agroindustrialization, globalization, and international development: The environmental implications. *Environment and Development Economics,* **6,** 419–433.

Bass, B. and J.R. Brook, 1997: Downscaling procedures as a tool for integration of multiple air issues. *Environmental Monitoring and Assessment,* **46,** 152–174.

Bauer, B.O., J.A. Winkler, and T.T. Veblen, 1999: Afterword: A shoe for all occasions or shoes for every occasion: Methodological diversity, normative fashions, and metaphysical unity in physical geography. *Annals of the Association of American Geographers,* **89(4),** 771–778.

Beierle, T.C. and J. Cayford, 2002: *Democracy in Practice: Public Participation in Environmental Decisions.* Resources for the Future, Washington, DC, 160 pp.

Belward, A., 1996: *The IGBP-DIS global 1 km land cover data set "DISCover" - Proposal and implementation plans.* Report of the Land Cover Working Group of the IGBP-DIS. IGBP-DIS Working Paper No. 13, Stockholm.

Berkes, F., 2002: Cross-scale institutional linkages: Perspectives from the bottom up. In: *The Drama of the Commons,* E. Ostrom, T. Dietz, N. Dolšak, P.C. Stern, S. Stonich, and E.U. Weber (eds.), National Academy Press, Washington, DC, 293–322.

Bernardo, J.M. and A.F.M. Smith, 2000: *Bayesian Theory.* Wiley, New York, NY.

Berry, B.J.L., 1991: *Long-Wave Rhythms in Economic Development and Political Behavior.* Johns Hopkins University, Baltimore, MD.

Berry, B.J.L., 2000: A pacemaker for the Long Wave. *Technological Forecasting and Social Change,* **63,** 1–23.

Binswager, H., 1989: *Brazilian Policies that Encourage Deforestation in the Amazon.* Environment Department Working Paper, World Bank, Washington, DC.

Bisonette, J.A. (ed.), 1997: *Wildlife and Landscape Ecology: Effects of Pattern and Scale.* Springer-Verlag, Berlin.

Blöschl, G., 1996: *Scale and Scaling in Hydrology.* Habilitationsschrift, Vienna Technical University, Vienna.

Blöschl, G. and M. Sivapalan, 1995: Scale issues in hydrological modelling: A review. *Hydrological Processes,* **9,** 251–290.

Braden, J.B. and C.D. Kolstad (eds.), 1991: *Measuring the Demand for Environmental Quality.* Contributions to Economic Analysis No. 198, North-Holland, Amsterdam.

Broecker, W.S., 1997: Thermohaline circulation, the Achilles heel of our climate system: Will man-made CO_2 upset the current balance? *Science*, **278**, 1582–1588.

Bromley, D., 1990: The ideology of efficiency: Searching for a theory of policy analysis. *Journal of Environmental Economics and Management*, **19**, 86–107.

Brooks, D., H. Pajuoja, T.J. Peck, B. Solberg, and P.A. Wardle, 1996: Long-term trends and prospects in world supply and demand for wood. In: *Long-Term Trends and Prospects in World Supply and Demand for Wood*, B. Solberg (ed.), European Forest Institute, Finland.

Buck, S.J., 1998: *The Global Commons: An Introduction*. Earthscan, London, 225 pp.

Bugmann, H., M. Lindner, P. Lasch, M. Flechsig, B. Ebert, and W. Cramer, 2000: Scaling issues in forest succession modeling. *Climatic Change*, **44**, 265–289.

Butler, C.D., 2000: Inequality, global change and the sustainability of civilisation. *Global Change and Human Health*, **1(2)**, 156–172.

Cairns, J., 1977: Quantification of biological integrity. In: *The Integrity of Water*, R.K. Ballentine and L.J. Guarraia (eds.), U.S. Environmental Protection Agency, Office of Water and Hazardous Materials, Washington, DC, 171–187.

Callicott, J.B., 1989: *In Defense of the Land Ethic: Essays in Environmental Philosophy*. State University of New York Press, Albany, NY.

Callicott, J.B., 1994: *Earth's Insights: A Multicultural Survey of Ecological Ethics from the Mediterranean Basin to the Australian Outback*. University of California Press, Berkeley, CA.

Campell, B. and M. Luckert (eds.), 2002: *Uncovering the Hidden Harvest: Valuation Methods for Woodland and Forest Resources*. Earthscan, London.

Campell, D.T., 1969: Reforms as Experiments. *American Psychologist*, **24**, 409–429.

Carney, D., (ed.), 1998: Sustainable rural livelihoods: What contributions can we make? Paper presented at the *Natural Resources Advisers' Conference*, July. Department for International Development, London, 213 pp.

Carpenter, S.R., 2002: Ecological futures: Building an ecology of the long now. *Ecology*, **83(8)**, 2069–2083.

Carpenter, S.R., 2003: *Regime Shifts in Lake Ecosystems: Pattern and Variation. Excellence in Ecology Series*, Ecology Institute, Oldendorf/Luhe, Germany.

Cash, D.W. and S.C. Moser, 1998: Cross-scale interactions in assessments, information systems, and decision-making. In: *Critical Evaluation of Global Environmental Assessments*, Global Environmental Assessment Project, Harvard University, Cambridge, MA.

Casman, E.A., M.G. Morgan, and H. Dowlatabadi, 1999: Mixed levels of uncertainty in complex policy models. *Risk Analysis*, **19(1)**, 33–42.

Castro, R., F. Tattenbach, L. Gámez, and N. Olson, 1998: *The Costa Rican Experience with Market Instruments to Mitigate Climate Change and Conserve Biodiversity*. Fundecor and MINAE, San José, Costa Rica.

Chambers, R., 1997a: Responsible well-being — A personal agenda for development. *World Development*, **25(11)**, 1743–1754.

Chambers, R., 1997b: *Whose Reality Counts? Putting the First Last*. Intermediate Technology Publications, London, 297 pp.

Chapple, C.K., 1986: Non-injury to animals: Jaina and Buddhist perspectives. In: *Animal Sacrifices: Religious Perspectives on the Use of Animals in Science*, T. Regan (ed.), Temple University Press, Philadelphia, PA.

Chertow, M., 2001: The IPAT equation and its variants: Changing views of technology and environmental impact. *Journal of Industrial Ecology*, **4**, 13–29.

Chess, C. and K. Purcell, 1999: Public participation and the environment: Do we know what works? *Environmental Science and Technology*, **33**, 2685–2692.

Chess, C., T. Dietz, and M. Shannon, 1998: Who should deliberate when? *Human Ecology Review*, **5**, 45–48.

Chopra, K. and S.C. Gulati, 2001: *Migration and Common Property Resources: A Study in the Arid and Semi-arid Regions of India*. Sage Publications, New Delhi and London.

Chopra, K. and P. Dasgupta, 2002: *Common Pool Resources in India: Evidence, Significance and New Management Initiatives*. Report for DFID project on policy implications of common pool resource knowledge in India, Tanzania, and Zimbabwe. Available at http://www.-cpr.geog.cam.ac.uk.

Chopra, K. and A. Duraiappah, in press: Operationalising capabilities in a segmented society: The role of institutions. In: *Operationalising Capabilities*, F. Comim (ed.), Cambridge University Press, Cambridge. (Paper presented at the *Conference on Justice and Poverty: Examining Sen's Capability Approach*, June 2001. Cambridge University, Cambridge. Available at http://www.st-edmunds.cam.ac.uk/vhi/sen/program1.shtml.)

Chopra, K., G.K. Kadekodi, and M.N. Murty, 1990: *Participatory Development and Common Property Resources*. Sage Publications, New Delhi and London, 163 pp.

Clark, J.S., S.R. Carpenter, M. Barber, S. Collins, A. Dobson, J.A. Foley, D.M. Lodge, M. Pascual, R. Pielke, W. Pizer, C. Pringle, W.V. Reid, K.A. Rose, O.E. Sala, W.H. Schlesinger, D. Wall, and D. Wear, 2000: Ecological forecasting: An emerging imperative. *Science*, **293**, 657–660.

Clark, W.C., 1985: Scales of climate impacts. *Climatic Change*, **7**, 5–27.

Clark, W.C. and N.M. Dickson, 1999: The global environmental assessment project: Learning from efforts to link science and policy in an interdependent world. *Acclimations*, **8**, 6–7.

Coe, M.T., 2000: Modeling terrestrial hydrological systems at the continental scale: Testing the accuracy of an atmospheric GCM. *Journal of Climate*, **13**, 686–704.

Contreras-Hermosilla, A., 2000: *The Underlying Causes of Forest Decline*. CIFOR Occasional Paper 30, Center for International Forestry Research, Bogor, Indonesia.

Cosgrove, W. and F. Rijsberman, 2000: *World Water Vision: Making Water Everybody's Business*. Earthscan, London.

Cosmides, L. and J. Tooby, 1996: Are humans good intuitive statisticians after all? Rethinking some conclusions from the literature on judgment under uncertainty. *Cognition*, **58**, 1–73.

Costanza, R., 2000: Societal goals and the valuation of ecosystem services. *Ecosystems*, **3**, 4–10.

Costanza, R. and T. Maxwell, 1994: Resolution and predictability: An approach to the scaling problem. *Landscape Ecology*, **9**, 47–57.

Costanza, R. and C. Folke, 1996: The structure and function of ecological systems in relation to property rights regimes. In: *Rights to Nature*, S. Hanna, C. Folke, and K.G. Maler (eds.), Island Press, Washington, DC, 13–34.

Costanza, R., B. Norton, and B. Haskell (eds.), 1992: *Ecosystem Health: New Goals for Environmental Management*. Island Press, Washington, DC.

Costanza, R., R. D'Arge, R.S. de Groot, S. Farber, M. Grasso, B. Hannon, K. Limburg, S. Naeem, R.V. O'Neill, J. Paruelo, R.G. Raskin, P. Sutton, and M. van den Belt, 1997: The value of the world's ecosystem services and natural capital. *Nature*, **387(6630)**, 253–260.

Cowling, R.M., P.J. Mustart, H. Laurie, and M.B. Richards, 1994: Species diversity: Functional diversity and functional redundancy in fynbos communities. *South African Journal of Science*, **90**, 333–337.

Cox, P.M., R.A. Betts, C.D. Jones, S.A. Spall, and I.J. Totterdell, 2000: Acceleration of global warming due to carbon-cycle feedbacks in a coupled climate model. *Nature*, **408(6809)**, 184–187.

Cruz, W. and R. Repetto, 1992: *The Environmental Effects of Stabilization and Structural Adjustment Programs: The Philippines Case*. World Resources Institute, Washington, DC, 90 pp.

Daily, G.C. (ed.), 1997a: *Nature's Services: Societal Dependence on Natural Systems*. Island Press, Washington, DC, 392 pp.

Daily, G.C., 1997b: Introduction: What are ecosystem services? In: *Nature's Services: Societal Dependence on Natural Ecosystems*, G.C. Daily (ed.), Island Press, Washington, DC, 1–10.

Daily, G.C. and K. Ellison, 2002: *The New Economy of Nature: The Quest to Make Conservation Profitable*. Island Press, Washington, DC.

Daily, G.C., T. Söderqvist, S. Aniyar, K. Arrow, P. Dasgupta, P.R. Ehrlich, C. Folke, A.M. Jansson, B.O. Jansson, N. Kautsky, S. Levin, J. Lubchenco, K.G. Mäler, D. Simpson, D. Starrett, D. Tilman, and B. Walker, 2000: The value of nature and the nature of value. *Science*, **289**, 395–396.

Dasgupta, P., 1996: The economics of the environment. *Proceedings of the British Academy*, **90**, 165–221.

Dasgupta, P., 2001: *Human Well-Being and the Natural Environment*. Oxford University Press, Oxford, 305 pp.

Davies, S., 1996: *Adaptable Livelihoods: Coping with Food Insecurity in the Malian Sahel*. MacMillan Press Ltd., London, 335 pp.

de Groot, R.S., M. Wilson, and R. Boumans, 2002: A typology for the description, classification, and valuation of ecosystem functions, goods and services. *Ecological Economics*, **41(3)**, 393–408.

de Leo, G.A. and S. Levin, 1997: The multifaceted aspects of ecosystem integrity. [online] *Conservation Ecology*, **1(1)**:3. Available at http://www.consecol.org/vol1/iss1/art3.

de Vries, B. and J. Goudsblom (eds.), 2002: *Mappae Mundi: Humans and their Habitats in a Long-term Socio-ecological Perspective*. Amsterdam University Press, Amsterdam.

Delcourt, H.R., P.A. Delcourt, and T.I. Webb, 1983: Dynamic plant ecology: The spectrum of vegetation change in space and time. *Quarternary Science Review*, **1**, 153–175.

Deutch, E., 1970: Vedanta and ecology. In: *Indian Philosophical Annual*, T.M.P. Mahadevan (ed.), University of Madras, India.

DFID, 1999: *Sustainable Livelihoods Guidance Sheets*. Department for International Development, London, 8 sections.

Dietz, T., 1994: What should we do? Human ecology and collective decision making. *Human Ecology Review*, **1**, 301–309.

Dietz, T., 2003: What is a good decision? *Human Ecology Review*, **10**, 60–67.

Dietz, T. and E.A. Rosa, 1994: Rethinking the environmental impacts of population, affluence and technology. *Human Ecology Review*, **1**, 277–300.

Dietz, T. and P.C. Stern, 1998: Science, values and biodiversity. *BioScience*, **48**, 441–444.

Dietz, T. and P.C. Stern (eds.), 2002: *New Tools for Environmental Protection: Education, Information and Voluntary Measures*. National Academy Press, Washington, DC, 356 pp.

Dietz, T. and E.A. Rosa, 2002: Human dimensions of global change. In: *Handbook of Environmental Sociology*, R.E. Dunlap and W. Michelson (eds.), Greenwood Press, Westport, CT.

Dietz, T., R.S. Frey, and E. Rosa, 2002a: Risk, technology and society. In: *Handbook of Environmental Sociology*, R.E. Dunlap and W. Michelson (eds.), Greenwood Press, Westport, CT, 562–629.

Dietz, T., E. Ostrom, N. Dolšak, and P.C. Stern, 2002b: The drama of the commons. In: *The Drama of the Commons*, E. Ostrom, T. Dietz, N. Dolšak, P.C. Stern, S. Stonich, and E.U. Weber (eds.), National Academy Press, Washington, DC, 3–35.

Dixon, J.A., L.F. Scura, R.A. Carpenter, and P.B. Sherman, 1994: *Economic Analysis of Environmental Impacts*. Earthscan, London.

Dollar, D. and P. Collier, 2001: *Globalization, Growth, and Poverty: Building an Inclusive World Economy*. Oxford University Press, Oxford.

Donner, S.D., M.T. Coe, J.D. Lenters, T.E. Twine, and J.A. Foley, 2002: Modeling the impact of hydrological changes on nitrate transport in the Mississippi River Basin from 1955–1994. *Global Biogeochemical Cycles*, DOI:10.1029/2001GB001396, August 7.

Downs, T.J., 2000: Changing the culture of underdevelopment and unsustainability. *Journal of Environmental Planning and Management*, **43(5)**, 601–621.

Drèze, J. and A. Sen, 2002: *India: Development and Participation*. Oxford University Press, Oxford, 532 pp.

Dukes, J.S. and H.A. Mooney, 1999: Does global change increase the success of biological invaders? *Trends in Ecology and Evolution*, **14**, 135–139.

Duraiappah, A., 1998: Poverty and environmental degradation: A review and analysis of the nexus. *World Development*, **26(12)**, 2169–2179.

Duraiappah, A., 2002: *Poverty and Ecosystems: A Conceptual Framework*. UNEP Division of Policy and Law Paper, United Nations Environment Programme, Nairobi, 49 pp.

Easterling, W.E., L.O. Mearns, and C. Hays, 2000: Comparison of agricultural impacts of climate change calculated from high and low resolution climate change scenarios: Part II. The effect of adaptations. *Climatic Change*, (accepted).

Easterling, W.E., A. Weiss, C. Hays, and L.O. Mearns, 1998: Spatial scales of climate information for simulating wheat and maize productivity: The case of the U.S. Great Plains. *Agricultural and Forest Meteorology*, **90**, 51–63.

Eckberg, D.L. and T.J. Blocker, 1989: Varieties of religious involvement and environmental concerns: Testing the Lynn White Thesis. *Journal for the Scientific Study of Religion*, **28**, 509–517.

Eckberg, D.L. and T.J. Blocker, 1996: Christianity, environmentalism, and the theoretical problem of fundamentalism. *Journal for the Scientific Study of Religion*, **35**, 343–355.

Edmonds, J., M. Wise, H. Pitcher, R. Richels, T. Wigley, and C. MacCracken, 1996: An integrated assessment of climate change and the accelerated introduction of advanced energy technologies: An application of MiniCAM 1.0. *Mitigation and Adaptation Strategies for Global Change*, **1(4)**, 311–339.

EEA, 2001: *Designing Effective Assessments: The Role of Participation, Science and Governance, and Focus*. Environmental Issue Report No. 26, European Environment Agency, Luxembourg, 24 pp.

Ehrenfeld, D. and P.J. Bently, 1985: Judaism and the practice of stewardship. *Judaism*, **34**, 301–311.

Ellis, F., 1998: Livelihood diversification and sustainable rural livelihoods. In: *Sustainable Rural Livelihoods: What Contribution Can We Make?*, D. Carney (ed.), Papers presented at the Natural Resources Advisers' Conference, July 1998. Department for International Development, London, 53–65.

Ensmiger, J., 1997: Changing property rights: Reconciling formal and informal rights to land in Africa. In: *The Frontiers of the New Institutional Economics*, J.N. Drobak and J.V.C. Nye (eds.), Academic Press, New York, NY, 374 pp.

Environment Canada, 1997: *The Canada Country Study: Climate Impacts and Adaptation*. Adaptation and Impacts Research Group, Downsview, Ontario, Canada.

FAO, 2000: *FAO Yearbook 2000: Fishery Statistics Commodities*. Vol. 91, Food and Agriculture Organization of the United Nations, Rome.

FAO, 2003: FAOSTAT Statistics Database. [online] Food and Agriculture Organization of the United Nations, Rome. Available at http://www.fao.org/FAOSTAT.

Farber, S.C., R. Constanza, and M.A. Wilson, 2002: Economic and ecological concepts for valuing ecosystem services. *Ecological Economics*, **41**, 375–392.

Faust, B.B. and R.C. Smardon, 2001: Introduction and overview: Environmental knowledge, rights, ethics: Co-managing with communities. *Environmental Science and Policy*, **4**, 147–151.

Fenwick, A., A.K. Cheesmond, and M.A. Amin, 1981: The role of field irrigation canals in the transmission of Schistosoma mansoni in the Gezira Scheme, Sudan. *Bulletin of the World Health Organization*, **59**, 777–786.

Fernandez, L., 1999: An analysis of economic incentives in wetland policies addressing biodiversity. *The Science of the Total Environment*, **240**, 107–122.

Finlayson, A.C., 1994: *A Sociological Analysis of Northern Cod Stock Assessments from 1977–1990*. Social and Economic Studies No. 52, Institute of Social and Economic Research, Memorial University of Newfoundland, St. John's, Canada.

Fiorino, D.J., 1990: Citizen participation and environmental risk: A survey of institutional mechanisms. *Science, Technology and Human Values*, **15**, 226–243.

Foley, J.A., S. Levis, M.H. Costa, W. Cramer, and D. Pollard, 2000: Incorporating dynamic vegetation cover within global climate models. *Ecological Applications*, **10(6)**, 1620–1632.

Foley, J.A., I.C. Prentice, N. Ramankutty, S. Levis, D. Pollard, S. Sitch, and A. Haxeltine, 1996: An integrated biosphere model of land surface processes, terrestrial carbon balance, and vegetation dynamics. *Global Biogeochemical Cycles*, **10(4)**, 603–628.

Freeman III, M., 1993: *The Measurement of Environmental and Resource Values: Theory and Methods*. Resources for the Future, Washington, DC.

Gallopin, G. and F. Rijsberman, 2000: Three global water scenarios. *International Journal of Water*, **1(1)**, 16–40.

Gallopin, G., A. Hammond, P. Raskin, and R.J. Swart, 1997: *Branch Points: Global Scenarios and Human Choice*. Stockholm Environment Institute, Stockholm.

Gardner, G.T. and P.C. Stern, 1995: *Environmental Problems and Human Behavior*. Allyn and Bacon, Needham Heights, MA.

Geist, H.J. and E.F. Lambin, 2002: Proximate causes and underlying driving forces of tropical deforestation. *BioScience*, **52(2)**, 143–150.

Gelman, A., J.B. Carlin, H.S. Stern, and D.B. Rubin, 1995: *Bayesian Data Analysis*. Chapman and Hall, London.

Giampetro, M., in press: Complexity and scales: The challenge for integrated assessment. In: *Scaling Issues in Integrated Assessment*, J. Rotmans and D. Rothman (eds.), Swets & Zeitlinger, Lisse, Netherlands.

Gibson, C.C., E. Ostrom, and T.K. Ahn, 2000: The concept of scale and human dimensions of global change: A survey. *Ecological Economics*, **32(2)**, 217–239.

Gilbert, A.J. and R. Janssen, 1998: Use of environmental functions to communicate the values of a mangrove ecosystem under different management. *Ecological Economics*, **25**, 323–346.

Gill, S.D., 1987: *Mother Earth: An American Story*. University of Chicago Press, Chicago.

Gleick, P., 2000: *The World's Water 2000–2001*. Island Press, Washington, DC, 315 pp.

Goma, H.C., K. Rahim, G. Nangendo, J. Riley, and A. Stein, 2001: Participatory studies for agro-ecosystem evaluation. *Agriculture, Ecosystems and Environment*, **87**, 179–190.

Goodchild, M.F. and D.A. Quattrochi, 1997: Scale, multiscaling, remote sensing and GIS. In: *Scale in Remote Sensing and GIS*, D.A. Quattrochi and M.F. Goodchild (eds.), Lewis Publishers, Boca Raton, FL.

Goulder, L. and D. Kennedy, 1997: Valuing ecosystem services: Philosophical bases and empirical methods. In: *Nature's Services: Societal Dependence on Natural Ecosystems*, G.C. Daily (ed.), Island Press, Washington, DC.

Grossman, G. and A. Krueger, 1995: Economic growth and the environment. *Quarterly Journal of Economics*, **110(2)**, 353–377.

Guagnano, G.A., P.C. Stern, and T. Dietz, 1995: Influences on attitude-behavior relationships: A natural experiment with curbside recycling. *Environment and Behavior*, **27**, 699–718.

Guard, M. and M. Masaiganah, 1997: Dynamite fishing in Southern Tanzania, geographical variation, intensity of use and possible solutions. *Marine Pollution Bulletin*, **34(10)**, 758–762.

Gunderson, L.H. and C.S. Holling (eds.), 2002: *Panarchy: Understanding Transformations in Human and Natural Systems*. Island Press, Washington, DC.

Gunderson, L.H., C.S. Holling, and S.S. Light, 1995a: Barriers broken and bridges rebuilt: A synthesis. In: *Barriers and Bridges to the Renewal of Ecosystems and Institutions*, L.H. Gunderson, C.S. Holling, and S.S. Light (eds.), Columbia University Press, New York, NY, 489–532.

Gunderson, L.H., C.S. Holling, and S.S. Light (eds.), 1995b: *Barriers and Bridges to the Renewal of Ecosystems and Institutions*. Columbia University Press, New York, NY.

Hamilton, K. and M. Clemens, 1999: Genuine savings rates in developing countries. *World Bank Economic Review*, **13(2)**, 333–356.

Hanemann, W.M., 1991: Willingness to pay and willingness to accept: How much can they differ? *American Economic Review*, **81(3)**, 635–647.

Hanemann, W.M., 1992: Preface. In: *Pricing the European Environment*, S. Navrud (ed.), Scandinavian University Press, Oslo.

Hardi, P. and T. Zdan (eds.), 1997: *Assessing Sustainable Development: Principles in Practice*. International Institute for Sustainable Development, Winnipeg.

Hartwick, J., 1994: National wealth and net national product. *Scandinavian Journal of Economics*, **99(2)**, 253–256.

Harvey, L.D.D., 1997: Upscaling in global change research. In: *Elements of Change 1997: Session One: Scaling from Site-Specific Observations to Global Model Grids*, S.J. Hassol and J. Katzenberger (eds.), Aspen Global Change Institute, Aspen, CO, 14–33.

Harvey, L.D.D., 2000: Upscaling in global research change. *Climatic Change*, **44**, 225–263.

Heal, G., 2000a: Valuing ecosystem services. *Ecosystems*, **3**, 24–30.

Heal, G., 2000b: *Nature and the Marketplace: Capturing the Value of Ecosystem Services.* Island Press, Washington, DC.

Hellier, A., A.C. Newton, and S.O. Gaona, 1999: Use of indigenous knowledge for rapidly assessing trends in biodiversity: A case study from Chiapas, Mexico. *Biodiversity and Conservation*, **8**, 869–889.

Helliwell, D.R., 1969: Valuation of wildlife resources. *Regional Studies*, **3**, 41–49.

Hemmati, M., 2001: *Multi-Stakeholder Processes: A Methodological Framework: Executive Summary.* UNED Forum, London.

Heywood, V.H. and R.T. Watson (eds.), 1995: *Global Biodiversity Assessment.* Cambridge University Press, Cambridge.

Homer-Dixon, T.F., 1994: Environmental scarcities and violent conflict: Evidence from cases. *International Security*, **19(1)**, 5–40.

Houghton, J.T., Y. Ding, D. Griggs, M. Noguer, P.J. van der Linden, X. Dai, K. Maskell, and C.A. Johnson (eds.), 2001: *Climate Change 2001: The Science of Climate Change.* Cambridge University Press, Cambridge.

Hufschmidt, M.M., D.E. James, A.D. Meister, B.T. Bower, and J.A. Dixon, 1983: *Environment, Natural Systems, and Development: An Economic Valuation Guide.* Johns Hopkins University Press, Baltimore, MD.

Hughes, J.D., 1983: *American Indian Ecology.* Texas Western Press, El Paso, TX.

ICSU, 2002a: *Science and Technology for Sustainable Development.* ICSU Series on Sustainable Development No. 9, International Council for Science, Paris, 30 pp.

ICSU, 2002b: *Making Science for Sustainable Development More Policy Relevant.* ICSU Series on Science for Sustainable Development No. 8, International Council for Science, Paris, 28 pp.

ICSU, 2002c: *Science, Traditional Knowledge and Sustainable Development.* ICSU Series on Science for Sustainable Development No. 4, International Council for Science, Paris, 24 pp.

Indian National Academy of Sciences, Chinese Academy of Sciences, and U.S. National Academy of Sciences, 2001: *Growing Populations, Changing Landscapes: Studies from India, China, and the United States.* National Academy Press, Washington, DC.

IPCC, 2000: Land Use, Land-Use Change, and Forestry. R.T. Watson, I. Noble, B. Bolin, N. Ravidranath, D. Verardo, and D. Dokken (eds.), Intergovernmental Panel on Climate Change, Cambridge University Press, Cambridge.

IPCC, 2002: *Climate Change 2001: Synthesis Report.* Cambridge University Press, Cambridge.

IUCN, UNEP, and WWF, 1980: *World Conservation Strategy.* World Conservation Union, United Nations Environment Programme, Word Wide Fund for Nature, Gland.

IUCN, UNEP, and WWF, 1991: *Caring for the Earth.* World Conservation Union, World Wide Fund for Nature, United Nations Environment Programme, Gland.

Jacobs, M., 1997: Environmental valuation, deliberative democracy and public decision-making. In: *Valuing Nature: Economics, Ethics and Environment*, J. Foster (ed.), Rutledge, London, 211–231.

Jaeger, C., O. Renn, E.A. Rosa, and T. Webler, 2001: *Risk, Uncertainty and Rational Action.* Earthscan, London, 320 pp.

Jaganathan, N.V., 1989: *Poverty, Public Policies and the Environment.* Working Paper No. 24, Environment Department, World Bank, Washington, DC.

Jarvis, P.G. and K.G. McNaughton, 1986: Stomatal control of transpiration: Scaling up from leaf to region. *Advances in Ecological Research*, **15**, 1–49.

Jazairy, I., M. Alamgir, and T. Panuccio, 1992: *The State of the World Rural Poverty: An Inquiry into its Causes and Consequences.* New York University Press, New York, NY.

Jepson, P., J.K. Jarvie, K. MacKinnon, and K.A. Monk, 2001: The end for Indonesia's lowland forests? *Science*, **292**, 859–861.

Jodha, N.S., 2001: *Life on the Edge: Sustaining Agriculture and Community Resources in Fragile Environments.* Oxford University Press, New Delhi, 317 pp.

Jones, C.G., J.H. Lawton, and M. Shachak, 1994: Organisms as ecosystem engineers. *Oikos*, **69**, 373–386.

Jordan, B., 1996: *A Theory of Poverty and Social Exclusion.* Polity Press, Cambridge, 276 pp.

Kahneman, D., P. Slovic, and A. Tversky (eds.), 1982: *Judgement under Uncertainty: Heuristics and Biases.* Cambridge University Press, Cambridge, 551 pp.

Kainuma, M., Y. Matsuoka, and T. Morita, 2003: *Climate Policy Assessment.* Springer, Tokyo, 402 pp.

Kalupahana, D., 1985: Toward a middle path of survival. In: *Nature in Asian Traditions of Thought*, J.B. Callicott and R.T. Ames (eds.), State University of New York Press, Albany, NY.

Kant, I., 1959 [1785]: *Foundations of the Metaphysics of Morals.* Bobbs Merrill, New York, NY.

Karr, J.R. and D.R. Dudley, 1981: Ecological perspective on water quality goals. *Environmental Management*, **5**, 55–68.

Kasperson, J.X., R.E. Kasperson, and B.L. Turner II, 1995: *Regions at Risk: Comparisons of Threatened Environments.* United Nations University Press, Tokyo.

Kates, R.W. and V. Haarmann, 1992: Where people live: Are the assumptions correct? *Environment*, **34**, 4–18.

Kates, R.W., T.J. Wilbanks, and R. Abler (eds.), 2003: *Global Change in Local Places: Estimating, Understanding, and Reducing Greenhouse Gases.* Cambridge University Press, Cambridge.

Kaufmann-Hayoz, R., C. Bättig, S. Bruppacher, R. Defila, A. Di Giulio, P. Flury-Kleubler, U. Friederich, M. Garbely, H. Gutscher, C. Jäggi, M. Jegen, H.J. Mosler, A. Müller, N. North, S. Ulli-Beer, and J. Wichtermann, 2001: A typology of tools for building sustainability strategies. In: *Changing Things – Moving People: Strategies for Promoting Sustainable Development at the Local Level*, R. Kaufmann and H. Gutscher (eds.), Birkhäuser, Basel, 33–107.

Keck, M.E. and K. Sikkink, 1999: Transnational advocacy networks in international and regional politics. *International Social Science Journal*, **51(1)**, 89–101.

Kempton, W., J.S. Boster, and J.A. Hartley, 1995: *Environmental Values in American Culture.* The MIT Press, Cambridge, MA.

Kenmore, P. and R. Krell, 1998: Global perspective and pollination in agriculture and agroecosystem management. Paper presented at the *International Workshop on the Conservation and Sustainable Use of Pollinators in Agriculture with Emphasis on Bees.* Food and Agriculture Organization of the United Nations, São Paulo, Brazil.

King, R.T., 1966: Wildlife and man. *NY Conservationist*, **20(6)**, 8–11.

Koziell, I., 1998: Biodiversity and sustainable rural livelihoods. In: *Sustainable rural livelihoods: What contributions can we make?*, D. Carney (ed.), Papers presented at the Natural Resources Advisers' Conference, July 1998. Department for International Development, London, 83–92.

Krech III, S., 1999: *The Ecological Indian: Myth and History.* W.W. Norton & Company, New York, NY, 320 pp.

Kremen, C., J.O. Niles, M.G. Dalton, G.C. Daily, P.R. Ehrlich, J.P. Fay, and D. Grewal, 2000: Economic incentives for rain forest conservation across scales. *Science*, **288**, 1828–1831.

Kucharik, C.J., J.A. Foley, C. Delire, V.A. Fisher, M.T. Coe, J.D. Lenters, C. Young-Molling, N. Ramankutty, J.M. Norman, and S.T. Gower, 2000: Testing the performance of a dynamic global ecosystem model: Water balance, carbon balance and vegetation structure. *Global Biogeochemical Cycles*, **14(3)**, 795–825.

Kuznets, S., 1979: *Growth, Population and Income Distribution: Selected Essays.* Nortan & Company, New York, NY.

Lambin, E.F., B.L. Turner II, H.J. Geist, S.B. Agbola, A. Angelsen, J.W. Bruce, O. Coomes, R. Dirzo, G. Fischer, C. Folke, P.S. George, K. Homewood, J. Imbernon, R. Leemans, X. Li, E.F. Moran, M. Mortimore, P.S. Ramakrishnan, M.B. Richards, H. Skånes, W.L. Steffen, G.D. Stone, U. Svedin, T.A. Veldkamp, C. Vogel, and J. Xu, 2001: The causes of land-use and land-cover change: Moving beyond the myths. *Global Environmental Change*, **11**, 261–269.

Lee, K.N., 1993: *Compass and Gyroscope: Integrating Science and Politics for the Environment.* Island Press, Washington, DC, 243 pp.

Leopold, A., 1949: *A Sand County Almanac.* Oxford University Press, New York, NY.

Levin, S.A., 1992: The problem of pattern and scale in ecology. *Ecology*, **73**, 1943–1967.

Lindeman, R.E., 1942: The trophic dynamic aspect of ecology. *Ecology*, **23**, 399–418.

Lindley, D.V., 1985: *Making Decisions.* Wiley, New York, NY.

Lobo, G., 2001: Ecosystem Functions Classification. [online] Cited September 2002. Available at http://gasa3.dcea.fct.unl.pt/ecoman/delphi/.

Longhurst, A.R., 1991: Role of the marine biosphere in the global carbon cycle. *Limnology and Oceanography*, **36**, 1507–1526.

Loreau, M., S. Naeem, and P. Inchausti (eds.), 2002: *Biodiversity and Ecosystem Functioning.* Oxford University Press, Oxford.

Lovell, C., A. Madondo, and P. Moriarty, 2002: The question of scale in integrated natural resource management. *Conservation Ecology*, **5(2)**, 25.

Ludwig, D., 2000: Limitations of economic valuation of ecosystem services. *Ecosystems*, **3**, 31–35.

Ludwig, D., M. Mangel, and B. Haddad, 2001: Science, conservation, and public policy. *Annual Review of Ecology and Systematics*, **32**, 481–517.

Markandya, A., 2001: Poverty alleviation and sustainable development: Implications for the management of natural capital. Paper presented at the *Workshop on Poverty and Sustainable Development.* International Institute for Sustainable Development, Ottawa, Canada.

Matthews, E., R. Payne, M. Rohweder, and S. Murray, 2000: *Pilot Analysis of Global Ecosystems: Forest Ecosystems.* World Resources Institute, Washington, DC, 90 pp.

McCarthy, J.J., N. Canziani, N. Leary, D.J. Dokken, and K.S. White (eds.), 2001: *Climate Change 2001: Impacts, Adaptation and Vulnerability.* Cambridge University Press, Cambridge.

McCay, B.J., 2002: Emergence of institutions for the commons: Contexts, situations and events. In: *The Drama of the Commons*, E. Ostrom, T. Dietz, N. Dolšak, P.C. Stern, S. Stonich, and E.U. Weber (eds.), National Academy Press, Washington, DC, 361–402.

McCay, B.J. and S. Jentoft, 1998: Market or community failure? Critical perspectives on common property research. *Human Organization*, **57**, 21–29.

McConnell, W., 2002: Madagascar: Emerald isle or paradise lost? *Environment*, **44(8)**, 10–22.

McMichael, A.J., 2001: *Human Frontiers, Environments and Disease: Past Patterns, Uncertain Futures.* Cambridge University Press, Cambridge, 413 pp.

Meadows, D.H., D.L. Meadows, J. Randers, and W. Behrens, 1972: *The Limits to Growth.* Universe Books, New York, NY.

Melillo, J.M., A.D. McGuire, D.W. Kicklighter, B. Moore III, C.J. Vörösmarty, and A.L. Schloss, 1993: Global climate change and terrestrial net primary production. *Nature,* **363,** 234–240.

Moberg, F. and C. Folke, 1999: Ecological goods and services of coral reef ecosystems. *Ecological Economics,* **29(2),** 215–233.

Morgan, M.G., 1998: Uncertainty analysis in risk assessment. *Human and Ecological Risk Assessment,* **4(1),** 25–39.

Morgan, M.G. and M. Henrion, 1990: *Uncertainty: A Guide for Dealing with Uncertainty in Quantitative Risk and Policy Analysis.* Cambridge University Press, Cambridge.

Moss, R.H. and S.H. Schneider, 2000: Uncertainties in the IPCC TAR: Recommendations to lead authors for more consistent assessment and reporting. In: *Guidance Papers on the Cross-Cutting Issues of the Third Assessment Report of the IPCC,* R. Pachauri, T. Taniguchi, and K. Tanaka (eds.), World Meteorological Organization, Geneva, 33–51.

Murray-Darling Basin Ministerial Council, 2001: *Integrated Catchment Management in the Murray-Darling Basin 2001–2010: Delivering a Sustainable Future.* Murray-Darling Basin Commission, Canberra, Australia.

Myers, N. and J. Kent, 2001: *Perverse Subsidies.* Island Press, Washington, DC.

Myers, R.A., J. Bridson, and N.J. Borrowman, 1995: *Summary of Worldwide Stock and Recruitment Data.* Canadian Technical Report of Fisheries and Aquatic Science No. 2024.

Naeem, S., 1998: Species redundancy and ecosystem reliability. *Conservation Biology,* **12,** 39–45.

Nakícenovíc, N., J. Alcamo, G. Davis, B. de Vries, J. Fenhann, S. Gaffin, K. Gregory, T. Grübler, T.Y. Jung, T. Kram, E. Emilio la Rovere, L. Michaelis, S. Mori, T. Morita, W. Pepper, H. Pitcher, L. Price, K. Riahi, A. Roehrl, H.H. Rogner, A. Sankovski, M.E. Schlesinger, P.R. Shukla, S. Smith, R.J. Swart, S. van Rooyen, N. Victor, and Z. Dadi, 2000: *Special Report on Emissions Scenarios.* Cambridge University Press, Cambridge.

Narayan, D., R. Chambers, M.K. Shah, and P. Petesch, 1999: *Global Synthesis: Consultations with the Poor.* World Bank, Washington, DC, 41 pp.

Narayan, D., R. Chambers, M.K. Shah, and P. Petesch, 2000: *Voices of the Poor: Crying Out for Change.* Oxford University Press, New York, 314 pp.

Navrud, S. and R.C. Ready (eds.), 2002: *Valuing Cultural Heritage: Applying Environmental Valuation Techniques to Historic Buildings, Monuments and Artifacts.* Edward Elgar, Cheltenham, UK.

Naylor, R.L., R.J. Goldburg, J.H. Primavera, N. Kautsky, M.C.M. Beveridge, J. Clay, C. Folke, J. Lubchenco, H.A. Mooney, and M. Troell, 2000: Effect of aquaculture on world fish supplies. *Nature,* **405,** 1017–1024.

Neffjes, K., 2000: *Environments and Livelihoods: Strategies for Sustainability.* Oxfam Publishing, Oxford, 277 pp.

Nelson, G.C. and J. Geoghegan, 2002: Modeling deforestation and land use change: Sparse data environments. *Agricultural Economics,* **27,** 201–216.

Norberg, J., 1999: Linking Nature's services to ecosystems: Some general ecological concepts. *Ecological Economics,* **29(2),** 183–202.

Nordstrom, H. and S. Vaughan, 1999: *Trade and Environment.* World Trade Organization, Geneva.

Nowak, D.J., 1994: Air pollution removal by Chicago's urban forest. In: *Chicago's Urban Forest Ecosystem: Results of the Chicago Urban Forest Climate Project*, E.G. McPherson, D.J. Nowak, and R.A. Rowntree (eds.). Gen Tech Report NE–186, U.S. Department of Agriculture, Forest Service, Northwestern Forest Experiment Station, Radnor, PA, 63–81.

NRC, 1999: *Our Common Journey: A Transition Toward Sustainability*. National Research Council, National Academy of Sciences, Washington, DC, 384 pp.

O'Connor, J., 1988: Capitalism, nature, socialism: A theoretical introduction. *Capitalism, Nature, Socialism*, **1**, 11–38.

Odin, S., 1991: The Japanese concept of nature in relation to the environmental ethics and conservation aesthetics of Aldo Leopold. *Environmental Ethics*, **13**, 345–360.

Odum, E., 1953: *Fundamentals of Ecology*. W.B. Saunders, Philadelphia, PA.

OECD InterFutures Study Team, 1979: *Mastering the Probable and Managing the Unpredictable*. Organisation for Economic Co-operation and Development and International Energy Agency, Paris.

Olson, D.M. and E. Dinerstein, 1998: The Global 200: A representation approach to conserving the Earth's most biologically valuable ecoregions. *Conservation Biology*, **12**, 502–515.

Olson, M., 1965: *The Logic of Collective Action: Public Goods and the Theory of Groups*. Harvard University Press, Cambridge, MA.

O'Neill, R.V., 1988: Hierarchy theory and global change. In: *Scales and Global Change*, T.R. Rosswall, G. Woodmansee, and P.G. Risser (eds.), New York, NY, John Wiley & Sons, 29–45.

O'Neill, R.V. and A.W. King, 1998: Homage to St. Michael: Or why are there so many books on scale? In: *Ecological Scale: Theory and Applications*, D.L. Peterson and V.T. Parker (eds.), Columbia University Press, New York, NY, 3–15.

Ong, J.E., 2002: The hidden costs of mangrove services: Use of mangroves for shrimp aquaculture. Paper presented at the *International Science Roundtable for the Media*, June, Joint event of ICSU, IGBP, IHDP, WCRP, DIVERSITAS, START. Bali, Indonesia. Available at http://www.igbp.kva.se/prepcom4/summary_ong.html.

Orkin, S.H. and S.J. Morrison, 2002: Biomedicine: Stem cell competition. *Nature*, **418**, 25–27.

Ostrom, E., 1990: *Governing the Commons: The Evolution of Institutions for Collective Action*. Cambridge University Press, Cambridge, 279 pp.

Ostrom, E., J. Burger, C.B. Field, R.B. Norgaard, and D. Policansky, 1999: Revisiting the commons: Local lessons, global challenges. *Science*, **284**, 278–282.

Ostrom, E., T. Dietz, N. Dolšak, P.C. Stern, S. Stonich, and E.U. Weber (eds.), 2002: *The Drama of the Commons*. National Academy Press, Washington, DC, 534 pp.

Paci, C., A. Tobin, and P. Robb, 2002: Reconsidering the Canadian Environmental Impact Assessment Act: A place for traditional environmental knowledge. *Environmental Impact Assessment Review*, **21(2)**, 111–127.

Pagiola, S., 1996: *Economic Analysis of Investments in Cultural Heritage: Insights from Environmental Economics*. World Bank, Washington, DC.

Pagiola, S., G. Acharya, and J.A. Dixon, in press: *Economic Analysis of Environmental Impacts*. Earthscan, London.

Palloni, A., 1994: The relation between population and deforestation: Methods for drawing causal inferences from macro and micro studies. In: *Population and Environment: Rethinking the Debate*, A. Lourdes, M.P. Stone, and D.C. Major (eds.), Westview, Boulder, CO.

Parmesan, C. and G. Yohe, 2003: A globally coherent fingerprint of climate change impacts across natural systems. *Nature*, **421**, 37–42.

Pauly, D., V. Christensen, J. Dalsgaard, R. Froese, and F.C. Torres Jr., 1998: Fishing down marine food webs. *Science*, **279**, 860–863.

Pearce, D.W. and J.W. Warford, 1993: *World Without End: Economics, Environment, and Sustainable Development*. Oxford University Press, Oxford.

Perrow, C., 1984: *Normal Accidents: Living with High Risk Technologies*. Basic Books, New York, NY, 386 pp.

Peterson, D.L. and V.T. Parker (eds.), 1998: *Ecological Scale: Theory and Application*. Columbia University Press, New York.

Peterson, G., 2000: Scaling ecological dynamics: Self-organization, hierarchical structure, and ecological resilience. *Climatic Change*, **44**, 291–309.

Petkova, E., C. Maurer, N. Henninger, F. Irwin, J. Coyle, and G. Hoff, 2002: *Closing the Gap: Information, Participation, and Justice in Decision-making for the Environment*. World Resources Institute, Washington, DC, 157 pp.

Petschel-Held, G., A. Block, M. Cassel-Gintz, J. Kropp, M. Lüdeke, O. Moldehauer, F. Reusswig, and H.J. Schellnhuber, 1999: Syndromes of global change: A qualitative modelling approach to assist global environmental management. *Environmental Modelling and Assessment*, **4**, 295–314.

Pimentel, D. and C. Wilson, 1997: Economics and environmental benefits of biodiversity. *BioScience*, **47(11)**, 747–758.

Pinstrup-Andersen, P., R. Pandya-Lorch, and M.W. Rosegrant, 1997: *The World Food Situation: Recent Developments, Emerging Issues and Long-Term Prospects*. International Food Policy Research Institute, Washington, DC.

Power, M.E., D. Tilman, J.A. Estes, B.A. Menge, W.J. Bond, S. Mills, G.C. Daily, J.C. Castilla, J. Lubchenco, and R.T. Paine, 1996: Challenges in the quest for keystones. *BioScience*, **46**, 609–620.

Pratt, J.W., H. Raiffa, and R. Schlaifer, 1995: *Introduction to Statistical Decision Theory*. The MIT Press, Cambridge, MA.

Prentice, I.C., W. Cramer, S.P. Harrison, R. Leemans, R.A. Monserud, and A.M. Solomon, 1992: A global biome model based on plant physiology and dominance, soil properties and climate. *Journal of Biogeography*, **19**, 117–134.

Prescott-Allen, R., 2001: *The Wellbeing of Nations: A Country-by-Country Index of Quality of Life and the Environment*. Island Press, Washington, DC, 342 pp.

Pritchard, L., C. Folke, and L. Gunderson, 2000: Valuation of ecosystem services in institutional context. *Ecosystems*, **3**, 31–35.

Randall, A., 1998: What mainstream economists have to say about the value of biodiversity. In: *Biodiversity*, E.O. Wilson (ed.), National Academy Press, Washington, DC.

Rapport, D.J., C. Gaudet, and P. Calow (eds.), 1995: *Evaluating and Monitoring the Health of Large-Scale Ecosystems*. Springer-Verlag, Heidelberg, Germany.

Rapport, D.J., G. Bohm, D. Buckingham, J. Cairns, Jr., R. Costanza, J.R. Karr, H.A.M. de Kruijf, R. Levins, A.J. McMichael, N.O. Nielsen, and W.G. Whitford, 1999: Ecosystem health: The concept, the ISEH, and the important tasks ahead. *Ecosystem Health*, **5**, 82–90.

Raskin, P., G. Gallopin, P. Gutman, A. Hammond, and R.J. Swart, 1998: *Bending the Curve: Toward Global Sustainability*. Stockholm Environment Institute, Boston, MA.

Raskin, P., T. Banuri, G. Gallopin, P. Gutman, A. Hammond, R.W. Kates, and R.J. Swart, 2002: *Great Transition: The Promise and Lure of Times Ahead*. Stockholm Environment Institute, Boston, MA.

Rastetter, E.B., A.W. King, B.J. Cosby, G.M. Hornberger, R.V. O'Neill, and J.E. Hobbie, 1992: Aggregating fine-scale ecological knowledge to model coarser-scale attributes of ecosystems. *Ecological Applications*, **2**, 55–70.

Rawls, J., 1971: *A Theory of Justice*. Harvard University Press, Cambridge, MA.

Redman, C.L., 1999: *Human Impact on Ancient Environments*. The University of Arizona Press, Tucson, AZ.

Regan, T., 1983: *The Case for Animal Rights*. University of California Press, Berkeley, CA.

Reid, W.V., 1996: Beyond protected areas: Changing perceptions of ecological management objectives. In: *Biodiversity in Managed Landscapes*, R. Szaro (ed.), Oxford University Press, Oxford, 442–453.

Reid, W.V., 2001: Capturing the value of ecosystem services to protect biodiversity. In: *Managing Human Dominated Ecosystems*, G. Chichilnisky, G.C. Daily, P. Ehrlich, G. Heal, and J.S. Miller (eds.). 84, Monographs in Systemic Botany from the Missouri Botanical Garden, St. Louis, MO.

Renn, O., T. Webler, and P. Wiedemann (eds.), 1995: *Fairness and Competence in Citizen Participation: Evaluating Models for Environmental Discourse*. Kluwer Academic Publishers, Dordrecht, The Netherlands.

Revenga, C., J. Brunner, N. Henninger, K. Kassem, and R. Payne, 2000: *Pilot Analysis of Global Ecosystems: Freshwater Systems*. World Resources Institute, Washington, DC, 83 pp.

Roberts, J.T. and P.E. Grimes, 1997: Carbon intensity and economic development 1962–1971: A brief exploration of the environmental Kuznets curve. *World Development*, **25**, 191–198.

Rolston III, H., 1994: *Conserving Natural Value*. Columbia University Press, New York, NY.

Root, T.L. and S.H. Schneider, 1995: Ecology and climate: Research strategies and implications. *Science*, **269**, 334–341.

Rosegrant, M.W., X. Cai, and S. Cline, 2002: *World Water and Food to 2025*. International Food Policy Institute, Washington, DC, 322 pp.

Rothschild, B.J., 1986: *Dynamics of Marine Fish Populations*. Harvard University Press, Cambridge, MA, 277 pp.

Rotmans, J. and D. Rothman (eds.), in press: *Scaling Issues in Integrated Assessment*. Swets & Zeitlinger, Lisse, Netherlands.

Rotmans, J., M. van Asselt, C. Anastasi, S. Greeuw, J. Mellors, S. Peters, D. Rothman, and N. Rijkens, 2000: Visions for a sustainable Europe. *Futures*, **32(9/10)**, 809–831.

Roy, A., 1999: *The Cost of Living: The Greater Common Good and the End of Imagination*. Flamingo, London, 161 pp.

Rudel, T. and J. Roper, 1997: The paths to rain forest destruction: Cross-national patterns of tropical deforestation, 1975–1990. *World Development*, **25**, 53–65.

Sagoff, M., 1988: *The Economy of the Earth*. Cambridge University Press, Cambridge.

Sagoff, M., 1998: Aggregation and deliberation in valuing environmental public goods: A look beyond contingent valuation. *Ecological Economics*, **24**, 213–230.

Sala, O.E., S.F. Stuart III, J.J. Armesto, E. Berlow, J. Bloomfield, R. Dirzo, S.E. Huber, L.F. Huenneke, R.B. Jackson, A. Kinzig, R. Leemans, D.M. Lodge, H.A. Mooney, M. Oesterheld, N.L. Poff, M.T. Sykes, B.H. Walker, M. Walker, and D.H. Wall, 2000: Biodiversity: Global biodiversity scenarios for the year 2100. *Science*, **287**, 1770–1774.

Scheffer, M., S.R. Carpenter, J.A. Foley, C. Folke, and B.H. Walker, 2001: Catastrophic shifts in ecosystems. *Science*, **413**, 591–596.

Schellnhuber, H.J. and V. Wenzel (eds.), 1998: *Earth System Science: Integrating Science for Sustainability*. Springer-Verlag, Heidelberg.

Schneider, S.H., B.L. Turner, and H. Morehouse Garriga, 1998: Imaginable surprise in global change science. *Journal of Risk Research*, **1(2)**, 165–185.

Schulze, E.D. and H.A. Mooney (eds.), 1993: *Biodiversity and Ecosystem Function*. Springer-Verlag, New York, NY.

Scoones, I., 1998: *Sustainable Rural Livelihoods: A Framework for Analysis*. Working Paper 72, Institute of Development Studies, University of Sussex, Brighton, UK, 21 pp.

Secretariat of the Convention on Biological Diversity, 2001: *Global Biodiversity Outlook*. United Nations Environment Programme and Convention on Biological Diversity, Montreal, 282 pp.

Sen, A.K., 1987: *On Ethics and Economics*. Basil Blackwell, Ltd., Oxford.

Sen, A.K., 1999: *Development as Freedom*. Oxford University Press, Oxford, 336 pp.

Sherman, K. and A.M. Duda, 1999: Large marine ecosystems: An emerging paradigm for fisheries sustainability. *Fisheries*, **24**, 15–26.

Shiklomanov, I.A., 1997: *Comprehensive Assessment of the Freshwater Resources of the World: Assessment of Water Resources and Water Availability in the World*. World Meteorological Organization and Stockholm Environment Institute, Stockholm.

Shogren, J. and J. Hayes, 1997: Resolving differences in willingness to pay and willingness to accept: A reply. *American Economic Review*, **87**, 241–244.

Simon, H.A., 1962: The architecture of complexity. *Proceedings of the American Philosophical Society*, **106**, 467–482.

Simon, H.A., 1974: The organization of complex systems. In: *Hierarchy Theory: The Challenge of Complex Systems*, H.H. Pattee (ed.), George Braziller, New York, NY.

Slocum, R., L. Wichhart, D. Rocheleau, and B. Thomas-Slayter, 1995: *Power, Process and Participation: Tools for Change*. Intermediate Technologies Publications, London.

SRES (Special Report on Emission Scenarios), 2000: Summary for Policymakers. [online] Working Group III of the Intergovernmental Panel on Climate Change. Available at http://www.ipcc.ch/pub/SPM_SRES.pdf.

Stafford-Smith, D.M. and J.F. Reynolds (eds.), 2002: *Integrated Assessment and Desertification*. Dahlem University Press, Berlin.

Stanner, W.E.H., 1979: *The White Man Got No Dreaming*. Australian University Press, Canberra.

Stern, D.I., 1998: Progess on the environmental Kuznets curve? *Environment and Development Economics*, **3**, 173–196.

Stern, P.C. and H. Fineberg (eds.), 1996: *Understanding Risk: Informing Decisions in a Democratic Society*. National Academy Press, Washington, DC.

Stern, P.C., T. Dietz, N. Dolšak, E. Ostrom, and S. Stonich, 2002: Knowledge and questions after fifteen years of research. In: *The Drama of the Commons*, E. Ostrom, T. Dietz, N. Dolšak, P.C. Stern, S. Stonich, and E.U. Weber (eds.), National Academy Press, Washington, DC, 443–490.

Tansley, A.G., 1935: The use and abuse of vegetational terms and concepts. *Ecology*, **16**, 284–307.

Taylor, P., 1986: *Respect for Nature*. Princeton University Press, Princeton, NJ.

Toth, F.L., 2000: Decision analysis frameworks in TAR. In: *Cross Cutting Issues Guidance Papers*, R. Pachauri, T. Taniguchi, and K. Tanaka (eds.), Intergovernmental Panel on Climate Change, Geneva, 53–68.

Tu, W.M., 1985: The continuity of being: Chinese visions of nature. In: *Nature in Asian Traditions of Thought*, J.B. Callicott and R.T. Ames (eds.), State University of New York Press, Albany, NY.

Turner II, B.L., D.L. Skole, S. Sanderson, G. Fischer, L.O. Fresco, and R. Leemans, 1995: *Land-Use and Land-Cover Change: Science/Research Plan*. IGBP Report No. 35 and HDP Report No. 7, International Geosphere-Biosphere Programme and the Human Dimensions of Global Environmental Change Programme, Stockholm.

Turner II, B.L., R.E. Kasperson, W.B. Meyer, K.M. Dow, D. Golding, J.X. Kasperson, R.C. Mitchell, and S.J. Ratick, 1990: Two types of global environmental change: Definitional and spatial scale issues in their human dimensions. *Global Environmental Change*, **1(1)**, 14–22.

Turner, M.G. and V.H. Dale, 1998: Comparing large, infrequent disturbances: What have we learned? *Ecosystems*, **1**, 493–496.

U.S. Census Bureau, 2002: International Data Base. 10 October. [online] U.S. Census Bureau, U.S. Department of Commerce. Available at http://www.census.gov/ipc/www/idbnew.html.

U.S. EPA Science Advisory Board, 2000: *Toward Integrated Environmental Decision-Making*. EPA-SAB-EC-00-011, United States Environmental Protection Agency, August, 46 pp.

U.S. Fish and Wildlife Service, 1999: *U.S. Fish and Wildlife Service Approves International Paper's Red Cockaded Woodpecker Habitat Management Plan*. News Release, February 18.

U.S. National Research Council, 1999: *Perspectives on Biodiversity: Valuing Its Role in an Everchanging World*. National Academy Press, Washington, DC.

UN Population Division, 2001: *World Population Prospects: The 2000 Revision*. ESA/P/WP 165, Department of Economic and Social Affairs, United Nations, New York, NY.

UN Population Division, 2002: *World Population Prospects: The 2001 Revision*. ST/ESA/SER.A/216, Department of Economic and Social Affairs, United Nations, New York, NY.

UNDP, 1998: *Human Development Report 1998*. United Nations Development Programme, New York, NY.

UNEP, 2002: *Global Environmental Outlook 2002*. United Nations Environment Programme, Nairobi.

UNFPA, 2002: *The State of World Population 2001*. United Nations Population Fund, New York, NY.

United Nations, 1992: *Rio Declaration on Environment and Development*. United Nations, New York, NY.

van Beers, C. and A.P.G. de Moor, 2001: *Public Subsidies and Policy Failures: How Subsidies Distort the Natural Environment, Equity and Trade and How to Reform Them*. Edward Elgar, Cheltenham, UK, 142 pp.

Vayda, A.P., 1988: Actions and consequences as objects of explanation in human ecology. In: *Human Ecology: Research and Applications*, R.J. Borden, J. Jacobs, and G.L. Young (eds.), Society for Human Ecology, College Park, MA, 9–18.

Victor, D.G., K. Raustiala, and E.B. Skolnikoff (eds.), 1998: *The Implementation and Effectiveness of International Environmental Commitments: Theory and Practice*. The MIT Press, Cambridge, MA, 737 pp.

Vitousek, P.M., J. Aber, R.W. Howarth, G.E. Likens, P.A. Matson, D.W. Schindler, W.H. Schlesinger, and D. Tilman, 1997: Human alteration of the global nitrogen cycle: Causes and consequences. *Ecological Applications*, **7**, 737–750.

Vörösmarty, C.J., B.I. Moore, A.L. Grace, M.P. Gildea, J.M. Melillo, B.J. Peterson, E.B. Rastetter, and P.A. Steudler, 1989: Continental scale models of water balance and fluvial transport: An application to South America. *Global Biogeochemical Cycles*, **3**, 241–265.

Wack, P., 1985: Scenarios: Shooting the rapids. *Harvard Business Review*, **64**, 135–150.

Wackernagel, M. and W. Rees, 1995: *Our Ecological Footprint: Reducing Human Impact on Earth*. New Society Publishers, Gabriola Island, BC, 160 pp.

Waggoner, P.E. and J.H. Ausubel, 2002: A framework for sustainability science: A renovated IPAT identity. *Proceedings of the National Academy of Sciences*, **99**, 7860–7865.

Walker, B.H., 1992: Biological diversity and ecological redundancy. *Conservation Biology*, **6**, 18–23.

Wall, D., C. Bock, T. Dietz, P. Hagenstein, A. Krzysik, R. Paine, S. Pimm, A. Randall, W.V. Reid, M. Sagoff, W. Schultze, D. Toweill, P. Vitousek, and D. Wake, 1999: *Perspectives on Biodiversity: Valuing its Role in an Ever-changing World*. National Research Council, National Academy Press, Washington, DC.

Walters, C., V. Christensen, and D. Pauly, 1997: Structuring dynamic models of exploited ecosystems from trophic mass-balance assessments. *Reviews in Fish Biology and Fisheries*, **7(2)**, 139–172.

Wang, G. and E.A.B. Eltahir, 2000: Ecosystem dynamics and the Sahel drought. *Geophysical Research Letters*, **27**, 795–798.

Watson, R.T., J.A. Dixon, S.P. Hamburg, A.C. Janetos, and R.H. Moss, 1998: *Protecting Our Planet — Securing Our Future*. United Nations Environment Programme, U.S. National Aeronautics and Space Administration, World Bank, Washington, DC.

WBCSD, 1997: Exploring Sustainable Development: Summary Brochure. [online] World Business Council for Sustainable Development. Available at http://www.wbcsd.org/newscenter/reports/1997/exploringscenarios.pdf.

WCED, 1987: *Our Common Future: The Bruntland Report*. Oxford University Press from the World Commission on Environment and Development, New York, NY, 400 pp.

Webler, T., 1999: The craft and theory of public participation: A dialectical process. *Journal of Risk Research*, **2**, 55–71.

Weins, J.A., 1989: Spatial scaling in ecology. *Functional Ecology*, **3**, 385–397.

White, L.J., 1967: The historical roots of our ecological crisis. *Science*, **155**, 1203–1207.

White, R.P., S. Murray, and M. Rohweder, 2000: *Pilot Analysis of Global Ecosystems: Grassland Ecosystems*. World Resources Institute, Washington, DC, 89 pp.

WHO, 1997: *The World Health Report 1997: Conquering Suffering, Enriching Humanity*. World Health Organization, Geneva.

Wilbanks, T.J., in press: Geographic scaling issues in integrated assessments of climate change. In: *Scaling Issues in Integrated Assessment*, J. Rotmans and D. Rothman (eds.), Swets & Zeitlinger, Lisse, Netherlands.

Wilbanks, T.J. and R.W. Kates, 1999: Global change in local places: How scale matters. *Climatic Change*, **43**, 601–628.

Wilson, J., 2002: Scientific uncertainty, complex systems and the design of common pool institutions. In: *The Drama of the Commons*, E. Ostrom, T. Dietz, N. Dolšak, P.C. Stern, S. Stonich, and E.U. Weber (eds.), National Academy Press, Washington, DC, 327–359.

Wilson, M.A. and R.B. Howarth, 2002: Valuation techniques for achieving social fairness in the distribution of ecosystem services. *Ecological Economics*, **41**, 431–443.

Wood, S.K., K. Sebastian, and S.J. Scherr, 2000: *Pilot Analysis of Global Ecosystems: Agroecosystems.* International Food Policy Research Institute and World Resources Institute, Washington, DC, 110 pp.

World Bank, 1997: *Expanding the Measure of Wealth: Indicators of Environmentally Sustainable Development.* Environmentally Sustainable Development Studies and Monographs No. 17, World Bank, Washington, DC.

World Bank, 2001: *World Development Report 2000/2001: Attacking Poverty.* Oxford University Press, Oxford, 335 pp.

World Bank, 2002a: *World Development Indicators 2002.* World Bank, Washington, DC, 432 pp.

World Bank, 2002b: *World Development Report 2003: Sustainable Development in a Dynamic World: Transforming Institutions, Growth, and Quality of Life.* Oxford University Press, New York, NY, 272 pp.

World Commission on Dams, 2000: *Dams and Development: A New Framework for Decision-Making.* Earthscan, London, 404 pp.

WRI, UNDP, UNEP, and World Bank, 2000: *World Resources 2000–2001: People and Ecosystems: The Fraying Web of Life.* World Resources Institute, Washington, DC, 389 pp.

Wu, J. and O.L. Loucks, 1995: From balance of nature to hierarchical patch dynamics: A paradigm shift in ecology. *Quarterly Review of Biology*, **70**, 439–466.

Yachi, S. and M. Loreau, 1999: Biodiversity and ecosystem functioning in a fluctuating environment: The insurance hypothesis. *Proceedings of the National Academy of Sciences*, **96**, 1463–1468.

Yohe, G. and F.L. Toth, 2000: Adaptation and the guardrail approach to tolerable climate change. *Climatic Change*, **45**, 103–128.

York, R., E. Rosa, and T. Dietz, 2003: Footprints on the Earth: The environmental consequences of modernity. *American Sociological Review*, (in press).

Young, O.R., 1994: The problem of scale in human/environment relations. *Journal of Theoretical Politics*, **6**, 429–447.

Young, O.R., 2002: *The Institutional Dimensions of Environmental Change: Fit, Interplay and Scale.* The MIT Press, Cambridge, MA.

Yunus, M., 1998: Alleviating technology through poverty. *Science*, **282**, 409–410.

Zaidi, I.H., 1981: On the ethics of man's interaction with the environment: An Islamic approach. *Environmental Ethics*, **3(1)**, 35–47.

Zimov, S.A., V.I. Chuprynin, A.P. Oreshko, F.S. Chapin III, J.F. Reynolds, and M.C. Chapin, 1995: Steppe-tundra transition: A herbivore-driven biome shift at the end of the Pleistocene. *American Naturalist*, **146**, 765–794.

Index

Italic page numbers refer to figures, tables, and boxes.
Bold page numbers refer to the Summary.